20 019 546 46

AF598494

Thin Films

Ionized Physical Vapor Deposition

Volume 27

Recent volumes in this serial appear at the end of this volume

Thin Films

Ionized Physical Vapor Deposition

Edited by

Jeffrey A. Hopwood

Northeastern University
Boston, Massachusetts

VOLUME 27

ACADEMIC PRESS

A Harcourt Science and Technology Company

San Diego San Francisco New York Boston
London Sydney Tokyo

This book is printed on acid-free paper ∞

ACADEMIC PRESS
A Harcourt Science and Technology Company
515 B Street, Suite 1900, San Diego, CA 92101-4495, USA
http://www.apnet.com

Academic Press Limited
24–28 Oval Road, London NW1 7DX, UK
http://www.hbuk.co.uk/ap/

International Standard Serial Number: 1079-4050
International Standard Book Number: 0-12-533027-8

Printed in the United States of America
99 00 01 02 03 COB 9 8 7 6 5 4 3 2 1

Contents

The Role of Ionized Physical Vapor Deposition in Integrated Circuit Fabrication

Jeffrey A. Hopwood

High-Density Plasma Sources

Amy E. Wendt

Ionization by Radio Frequency Inductively Coupled Plasma

Steve Rossnagel

Ionization by Microwave Electron Cyclotron Resonance Plasma

William M. Holber

Ionized Hollow Cathode Magnetron Sputtering

Kwok F. Lai

Applications and Properties of Ionized Physical Vapor Deposition Films

John Forster

Plasma Physics

Jeffrey A. Hopwood

Numerical Modeling

Ming Li, Michael A. Vyvoda, and David B. Graves

Contributors

Editor: Jeffrey A. Hopwood, Electrical and Computer Engineering Dept., Northeastern University, Boston, MA

The Role of Ionized Physical Vapor Disposition in Integrated Circuit Fabrication: Jeffrey A. Hopwood, Electrical and Computer Engineering Dept., Northeastern University, Boston, MA 02115

High-Density Plasma Sources: Amy E. Wendt, Electrical and Computer Engineering Dept., University of Wisconsin-Madison, Madison, WI 53706

Ionization by Radio Frequency Inductively Coupled Plasma: Stephen M. Rossnagel, IBM, Yorktown Heights, NY 10598

Ionization by Microwave Electron Cyclotron Resonance Plasma: William M. Holber, Applied Science and Technology, Woburn, MA 01801

Ionized Hollow Cathode Magnetron Sputtering: Kwok F. Lai, Novellus Systems, San Jose, CA 95134

Applications and Properties of Ionized Physical Vapor Deposition Films: John C. Forster, San Francisco, CA 94103

Plasma Physics: Jeffrey A. Hopwood, Electrical and Computer Engineering Dept., Northeastern University, Boston, MA 02115

Numerical Modeling: Ming Li, Michael A. Vyvoda, David B. Graves, Chemical Engineering, University of California, Berkeley, CA 94720

Preface

I have a confession to make: I am the type of person who reads the preface of a book. Apparently so are you! Why do we do this? Probably many reasons exist, but a compelling personal reason is to get an inside look at the authors—to see something other than the impersonal wall of technical and scientific information that follows the preface. Let the voyeurism begin.

"*Magic.*" That is how Stephen Campbell begins the preface of his text on integrated circuit fabrication.[1] What other word better sums up technology in the 20th century? In my lifetime we have progressed from that miraculous "7 Transistor" radio—the pride and joy of my baseball-addicted grandfather—to multimillion-transistor video games played by my Zelda-addicted father-in-law! While there is plenty of credit to go around, the practitioners of integrated circuit fabrication have driven this revolution from its roots. Without their steady march up the exponential mountain called Moore's law, I would be writing this with pen and paper, my father-in-law would be driving my mother-in-law to distraction when it is too cold to golf, and my grandfather... well, he still has his radio.

When I was a graduate student, doomsayers wrung their hands over the prospects of fabricating a transistor with a minimum feature size below 1 μm. Currently, 0.25-μm devices are routinely produced. Lithography of very small features was a major problem then, and it is still one of the major challenges today. However, as the size of individual transistors has decreased, designers have incorporated more transistors in a single chip. The complexity of interconnecting millions of transistors in a precise, reliable, and electrically fast way has skyrocketed. The aspect ratio of the metal wires has been increased so that interconnecting lines exhibit a low electrical resistance while being packed as densely as possible. Instead of the broad, flat metal lines of a few years ago, modern integrated circuits (ICs) sport tall, thin wires.

Ionized physical vapor deposition (I-PVD) is the result of a marriage between a new and an old technology. Everyone relied on sputtering to deposit metal interconnects for years, but the sputtered flux is inherently noncollimated and could not be made to fill, or even line, high aspect ratio interconnect structures. High-density plasma is the new arrival, having only been used in manufacturing during the past few years. When these two are used in combination, the high-density plasma ionizes the sputtered metal atoms. These metal ions can be collimated easily by the thin plasma sheath that forms above a wafer. Sputtering was granted a new life as I-PVD.

Magic is never as impressive once the magician's secrets are revealed. Nonetheless, in this book seven magician/scientists have agreed to describe one small part of microelectronic prestidigitation called I-PVD. After an introductory chapter on the role that I-PVD plays in microelectronics, Chapter 2 details the basic plasma physics and technology of the most common plasma sources used in I-PVD. Chapters 3–5 describe three different approaches to ionizing sputtered metal atoms using RF inductively coupled plasma, microwave electron cyclotron resonance plasma, and hollow cathode magnetron, respectively. Chapter 6 focuses on the microelectronic applications of I-PVD and the properties of the deposited thin films. Chapters 7 and 8 discuss the physics of I-PVD and numerical modeling of these plasma reactors.

So who should read this book? Anyone involved in back-end-of-the-line semiconductor processing tools will find these topics pertinent to their work. The material, however, is aimed at a broader audience of professionals and students who are interested in microelectronic materials, plasma processing, or IC fabrication. Although the focus of the text is microelectronic fabrication, a range of interesting and possibly new materials can be deposited from the energetic ions created by I-PVD. Novel materials usually deposited by energetic processes such as vacuum arc or laser ablation can also be made using I-PVD. Due to its simplicity, I-PVD may provide cleaner, more uniform films at a lower cost. Although we are currently in the infancy of this technology, the future applications are remarkably promising.

Many manufacturers of semiconductor equipment currently support a development program for I-PVD and a few offer a product. It is hoped that a person may intelligently choose a commercial system after reviewing this book. Alternatively, an experimentalist can quickly convert an existing sputtering system into an I-PVD reactor using the information presented here.

Jeffrey A. Hopwood

Reference

1. S. A. Campbell, *The Science and Engineering of Microelectronic Fabrication* (New York: Oxford University Press, 1996), vii.

Thin Films

Ionized Physical Vapor Deposition

Volume 27

The Role of Ionized Physical Vapor Deposition in Integrated Circuit Fabrication

JEFFREY A. HOPWOOD

Northeastern University, Boston, Massachusetts

I. Overview of Ionized Physical Vapor Deposition

Many sources used for the deposition of thin films produce ions. Therefore, the natural question is, What specifically distinguishes ionized physical vapor deposition (I-PVD) from other deposition techniques? By definition, I-PVD is a deposition process in which the depositing species are initially vaporized by physical mechanisms. In addition, the flux of depositing species must be composed of $>50\%$ ions. Ion plating is a well-known technique that is often confused with I-PVD. Ion plating, however, only ionizes a small fraction of the depositing species using a weak glow discharge ($n_e \sim 10^{10}\,\mathrm{cm}^{-3}$). I-PVD, by contrast, aggressively ionizes the depositing species using a high electron-density discharge ($n_e \sim 10^{12}\,\mathrm{cm}^{-3}$). Two other common deposition techniques using ions are known as ion beam-assisted deposition (IBAD) and ion-assisted deposition (IAD). These methods cobombard a film deposited from neutral atoms with inert gas ions. While the ratio of ion flux to the flux of depositing neutrals may be quite large in IBAD and IAD, the depositing species are primarily low-energy neutrals. The key distinction is that I-PVD seeks to deposit the film from ions.

Why are we concerned that the deposition flux primarily consists of ions? The main application for I-PVD is the formation of metal and nitride thin films into the deep, narrow trenches and vias that are found in modern integrated circuits (ICs). The motion of neutral atoms is difficult to control, but ions are easily collimated by the strong electric field region that forms near surfaces, such as wafers, immersed in a gas plasma. This region is known as the plasma sheath. As ions enter the sheath region, they are

Vol. 27
ISBN 0-12-533027-8

ISSN 1079-4050/00 $30.00

accelerated toward the wafer and collimated. In this way the ions may reach the bottom of deep, narrow trenches or vias that have been etched in the wafer. Most neutrals, on the other hand, will enter the vias at large angles and deposit predominantly on the sidewalls of these structures, leaving the bottom with very little film coverage.

In this book several I-PVD methods suitable for IC fabrication will be discussed. In general these tools create a metal vapor by sputtering a large-area target with argon ions. Sputtering is preferred because it is relatively easy to create the large, uniform flux of metal needed to coat 200- to 300-mm-diameter wafers. The metal vapor is then ionized as it passes through a region of dense argon/nitrogen plasma. The plasma may be created by radio frequency induction (see Chapters 3 and 6), microwave electron cyclotron resonance (see Chapter 4), or DC hollow cathode magnetron (see Chapter 5). Figure 1 shows a small laboratory I-PVD reactor made from a magnetron with a 75-mm aluminum sputter target and an inductively coupled plasma excited by a two-turn coil immersed in the vacuum chamber.

It is worth mentioning that this book does not attempt to exhaustively describe all methods of I-PVD. Filtered cathodic arc deposition, direct ion beam deposition, and some forms of laser ablation can all be used to grow thin films from ions. The stringent requirements of IC manufacturing, however, preclude these technologies from being used. Manufacturing requirements include very low particulate generation, broad area coverage, reliable operation in reactive environments, purity, and low cost. Developers

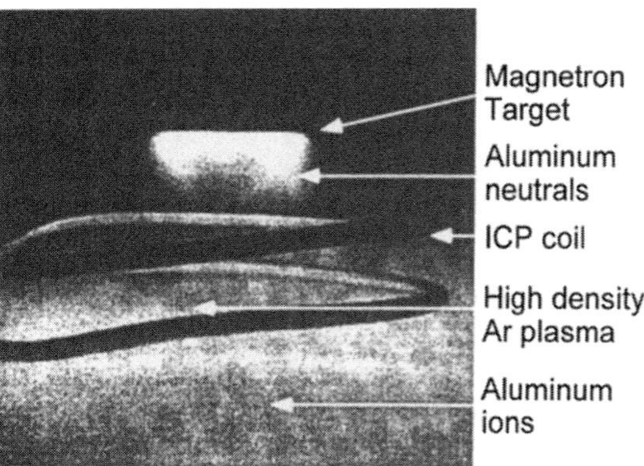

FIG. 1. A photograph of a lab-scale I-PVD system shows a 75-mm-diameter aluminum sputter target and an inductively coupled plasma used to ionize the aluminum atoms. The wafer (not shown) would be placed at the bottom of the figure in the region of high Al^+ concentration.

of filtered cathodic arcs, for example, have made great strides in reducing the particulate generation, but the macroparticle density is still too high for IC fabrication applications.

II. Trends in IC Fabrication

Cost and performance are the driving considerations in microelectronics. Performance can be separated into two areas: the number of functions available on a chip and the number of times per second these functions can be completed. The ability to make very small transistors currently allows electronic designers to use tens of millions of transistors in a single IC. Function, of course, depends on precisely how these transistors are interconnected. As one can well imagine, interconnecting a million transistors with microscopic "wires" is a daunting task. The connections are made by alternating layers of metal "lines" and insulators, called interlayer dielectrics (ILDs). Figure 2 provides a simplified cross-section view of an IC showing only a single transistor [gate (G), source (S), and drain (D) in a doped well] and three layers of metal lines. Where appropriate, metal lines are connected by a vertical "via." Vias that connect directly to the transistor are often distinguished by the name "contacts" because the requirements for making an ohmic contact to silicon are different than those for the subsequent layers.

Currently, about six layers of metal are used in microprocessor design. The number of layers is expected to increase as the number of transistors increases. It is easy to observe that the complexity, cost, and performance of the interconnect scheme rivals or even exceeds that of the transistors. In fact, the speed at which functions can be performed is now limited by the propagation of signals through the interconnect as much as by the speed of the transistors.

The propagation delay caused by the interconnect can be attributed to the product of the resistance of the conductors and the parasitic capacitance between the conductors (RC) as shown in Fig. 3. In order to prevent the capacitance from increasing to unusable levels, the thickness of the ILD cannot be substantially reduced in future generations of ICs. The diameter of the vias and contacts, however, will need to decrease as the transistor size is scaled down. As a result, the aspect ratio of the via (depth/diameter) is projected to become quite large.

The fabrication of the interconnect may follow many paths as described in Chapter 6. One of the most advanced methods is called *dual damascene*. This fabrication sequence allows the lines and vias to be fabricated simulta-

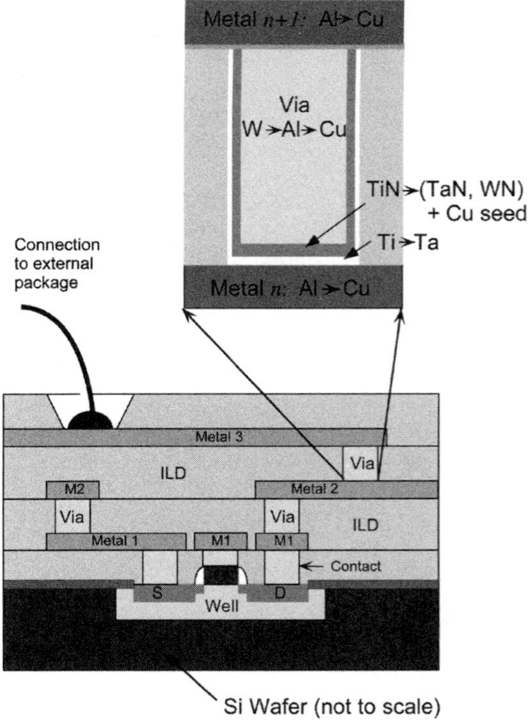

FIG. 2. A schematic cross-section sketch of an integrated circuit shows a single field effect transistor (FET) and three layers of electrical interconnections. The arrows indicate the predicted progression of materials used for the interconnects.

neously for each layer of the interconnect. One possible process flow for dual-damascene fabrication is shown in Fig. 4. First, an ILD is deposited and the via locations are etched (Fig. 4a). In a separate step, trenches are etched where the metal lines are desired (Fig. 4b). Next, it is necessary to deposit thin films of metal and metal nitrides using I-PVD or chemical vapor deposition (Fig. 4c). These layers serve as adhesion promoters, wetting surfaces, electrical contact enhancers, and barriers to diffusion. It may also be necessary to deposit a thin *seed layer* to initiate filling of the entire structure by electroplating in the following step (Fig. 4d). If electroplating is not used to fill, sputtering is used in combination with temperature-enhanced diffusion. The filled via/line structure is completed by planarizing the overburden of metal by chemical–mechanical polishing (CMP) (Fig. 4e). Subsequent layers are formed by depositing another ILD and repeating these steps.

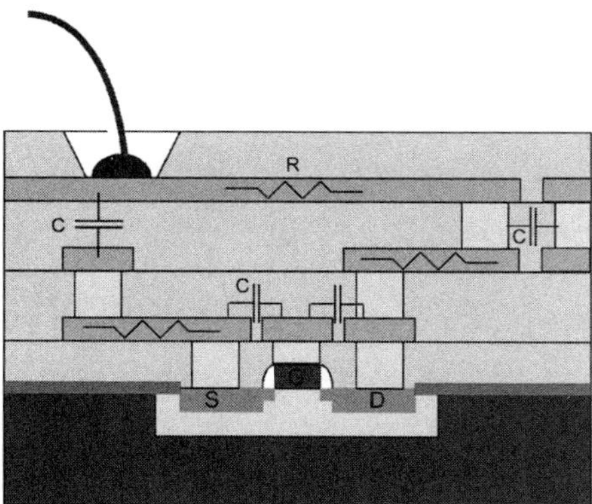

FIG. 3. The performance of a modern integrated circuit is limited by the propagation delay ($R \times C$) from one transistor to the next. As transistors become smaller, the parasitic capacitance of the FET is overwhelmed by the capacitance of the interconnect.

The "single" damascene process follows the same flow, except that the via is formed first by steps a, c, d, and e described for Fig. 4. Then another ILD is deposited and the line is created by steps b, c, d, and e described for Fig. 4. Clearly, the dual-damascene process uses fewer steps and has the potential

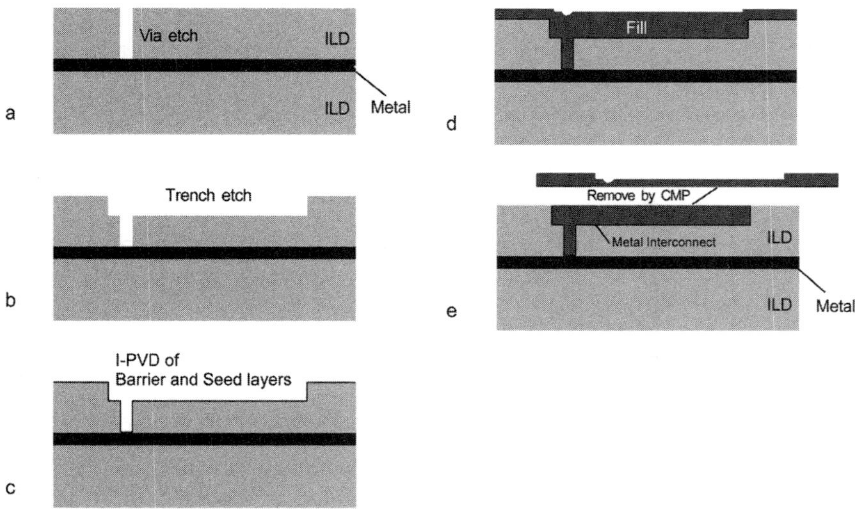

FIG. 4. The dual-damascene method shown here is an advanced method of fabricating IC interconnects. Among the technical challenges of dual-damascene fabrication, however, is lining and filling high aspect ratio features with metal and nitrides.

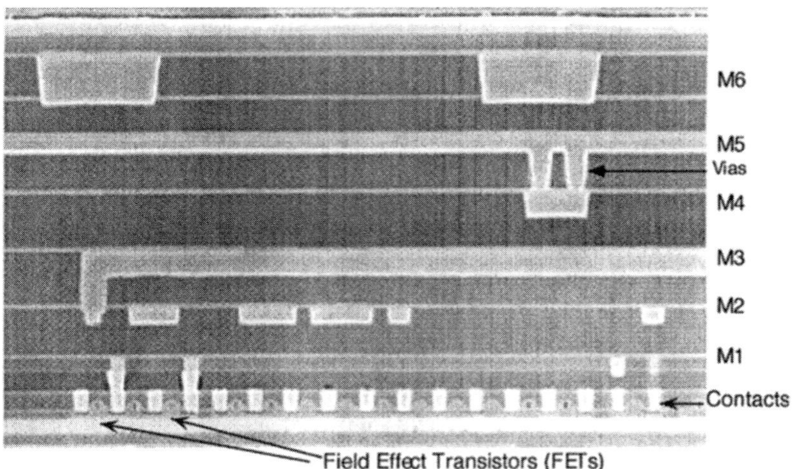

FIG. 5. A scanning electron micrograph of the cross section of an IBM PC 750 microprocessor shows six layers of copper interconnects. (Courtesy of Integrated Circuit Engineering Corporation, Scottsdale, AZ, USA).

to reduce costs. Among the problems of the dual-damascene process is the need to deposit films into the composite structure of the via and the trench simultaneously. The higher aspect ratio of the composite structure requires a deposition technique such as I-PVD because I-PVD is capable of depositing metal all the way to the bottom of the vias.

A final note on the materials used in IC interconnects is appropriate. Currently, aluminum, tungsten, and titanium are widely used, but copper is being introduced in an effort to reduce the resistance of the metal lines. Figure 5, for example, shows a cross section of the copper interconnects used by the IBM PC750 microprocessor. Compared to aluminum, copper is also less susceptible to electromigration failure that is caused by a high current density through the wire. Since copper is not readily etched anisotropically, the damascene method of patterning the vias and lines is used. In addition, copper does not form a self-limiting oxide and diffuses readily into the surrounding materials. Copper, therefore, must be fully encapsulated by new protective barrier layers. The anticipated progression of metal and barrier materials is indicated in Fig. 2. In this new materials set, I-PVD will likely be used to deposit Ta, TaN, and/or Cu seed layers.

III. Overview

The following chapters were written to be self-contained. This enables you to skip directly to the information that is needed. At the same time, for those who prefer to read cover to cover, the chapters are organized to flow in a logical progression. Chapter 2 discusses the basic operating principles of the various plasma generators used in I-PVD. Chapters 3–5 cover the operation and characteristics of three distinct I-PVD tools. The applications of I-PVD films to IC fabrication are discussed in Chapter 6. Chapters 7 and 8 describe the internal physical processes that occur in I-PVD and suggest how these principles can affect tool design and film quality.

Finally, although this book focuses on IC fabrication using I-PVD, there are many opportunities for creating novel materials from ion deposition. Chapter 4 references some of these interesting nonelectronic applications.

High-Density Plasma Sources

AMY E. WENDT

Department of Electrical and Computer Engineering and
The Center for Plasma-Aided Manufacturing,
University of Wisconsin — Madison, Madison, Wisconsin

I. Introduction

A. CHALLENGES FOR IONIZED PHYSICAL VAPOR DEPOSITION PLASMA SOURCE DESIGN

Ionized physical vapor deposition (I-PVD), in comparison with other process steps for ULSI (ultra large-scale integration) fabrication, presents some interesting and unique challenges in the selection of a plasma source. The material to be deposited starts out in solid form and is typically introduced into the gas phase through a plasma process known as magnetron sputtering. Magnetron sputtering has been a well-established and well-characterized method for depositing thin films since long before I-PVD was developed. The main difference between conventional sputtering and

Vol. 27
ISBN 0-12-533027-8
ISSN 1079-4050/00 $30.00

I-PVD is that in the latter, a significant fraction of the sputtered material must be ionized before depositing on the substrate. An I-PVD source differs from a conventional sputter source in that some means is required to enhance the ionization of the sputtered material. There are several approaches that have been explored, most of which involve a second plasma source in addition to the magnetron to ionize the sputtered material to be deposited.

B. SPUTTERING AND METAL IONIZATION IN THE SAME PROCESS TOOL

To achieve maximum control over the deposited films, the capability of a high degree of ionization of the sputtered material is required. Some ionization of sputtered material takes place even in conventional magnetron sputtering systems. As a sputtered atom passes through the magnetron plasma on its way to the substrate, it may become ionized through a collision with an energetic electron. The probability of ionization is limited both by the concentration of energetic electrons available for ionization and by the residence time of the sputtered atom in the plasma. The goal of all approaches to I-PVD is to provide a means of enhancing the degree of ionization of the sputtered material before it reaches the substrate.

C. THE NEED FOR HIGH-DENSITY PLASMA SOURCES

One general approach to enhancing the ionization of the sputtered material is to increase the distance between the sputter target and substrate in order to generate a "high-density plasma" in the space created. This increases the residence time for sputtered atoms both by increasing their path length to the substrate and by increasing the number of energetic electrons they encounter along that path. Increasing the path length between sputter target and substrate presents a trade-off in that the greater the distance, the greater the fraction of the sputtered material deposited on the sidewalls of the chamber and wasted. As far as the secondary plasma is concerned, however, the higher the plasma density, the higher the degree of ionization of sputtered material for a given set of conditions. Even when high-density plasmas are used, a significant flux of neutral atoms reaches the substrate in typical process tool configurations.

Several types of high-density plasma sources have been employed for I-PVD processes and will be described in this chapter. Most approaches to I-PVD use some form of DC magnetron discharge as a source of sputtered metal. Inductively coupled plasmas (ICP) and electron cyclotron resonance

(ECR) plasmas are used as secondary plasma sources to enhance the ionization of metal atoms sputtered from DC magnetron sources. Finally, the hollow cathode magnetron source, developed especially for I-PVD, combines sputtering and ionization in a single source.

II. DC Magnetron Discharges for Sputtering of Conducting Materials

Magnetron sputtering is used routinely in the deposition of thin metal films and constitutes one of the two major components of many I-PVD systems. A planar magnetron discharge is shown in Fig. 1. The magnetron is a magnetized high-density plasma that introduces metal atoms into the gas phase through sputtering. The sputter target serves both as the source of metal atoms to be deposited and as a cathode for the magnetron discharge. A DC voltage is typically used to drive an argon discharge for metal sputtering applications. A magnetic field is produced by magnets located behind the cathode so that magnetic field lines both enter and exit the plasma volume through the target surface. This magnetic field serves to confine the plasma near the target. A negative voltage of several hundred volts is applied to the cathode, and most of the voltage drop appears in a thin "sheath" between the cathode and the plasma. Ions accelerated across the sheath strike the target with energy of several hundred electron volts, enough to sputter atoms off the target surface. Many studies have contributed to the characterization and understanding of magnetron operation, and it remains a topic of current research.[1-11]

The magnetic field plays several roles in enhancing the operation of the magnetron sputter source. Plasma confinement by the magnetic field leads

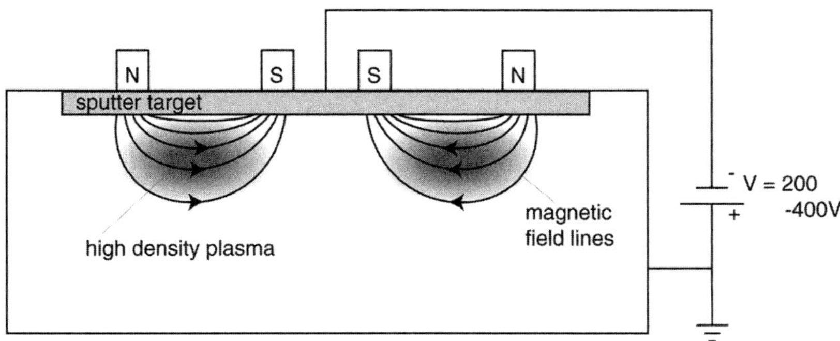

FIG. 1. Schematic diagram of planar magnetron sputtering system.

to a high plasma density close to the target, resulting in a higher sputter rate compared with that of a low-density, unmagnetized plasma. In addition, the magnetic confinement may be used to reduce exposure of semiconductor substrates to potentially damaging energetic electrons. Finally, the deposition rate at the substrate is enhanced because discharge operation can be extended to lower pressure when the magnetron magnetic field is used. Scattering of the sputtered metal atoms by the background gas (typically argon), which can lead to losses to the chamber walls and redeposition on the target, is minimized by operation at low pressure. Argon pressures on the order of 5 mTorr or lower are typical for conventional magnetron sputtering, but higher pressures are necessary for optimal I-PVD operation.

A. PRINCIPLE OF OPERATION

Ions reaching the sheath region are accelerated toward the cathode by the sheath electric field so that they strike the target surface with fairly high energy (several hundred electron volts). The impact of ions on the target has two consequences which are both critical to the magnetron sputter source operation. First, electrons are released from the target surface due to secondary electron emission by ion impact. The secondary emission coefficient, γ, is a function of both the target material and the identity and energy of the impinging ions. For argon ions impinging on a copper target under typical magnetron operating conditions, $\gamma \sim 0.1$. The electrons released through secondary emission are accelerated in the opposite direction through the sheath into the plasma. These electrons enter the plasma with an energy of several hundred electron volts and are trapped near the target by the magnetic field, as shown in Fig. 2. These electrons, through ionizing collisions with background gas atoms, are responsible for sustaining the magnetron discharge. Second, as a result of ion bombardment, metal atoms

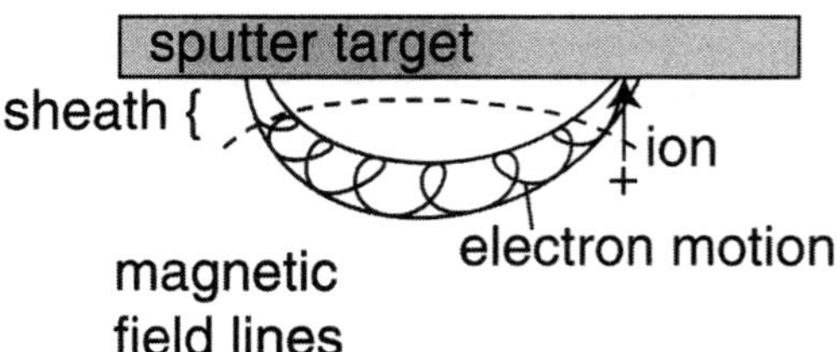

FIG. 2. Bombardment of the cathode surface by ions accelerated through the sheath results in emission of secondary electrons. These electrons reenter the plasma with substantial energy gained in the sheath electric field and are trapped in the magnetron magnetic field.

are knocked or "sputtered" off the target surface and enter the gas phase. The sputter yield (number of atoms sputtered per ion) depends both on target material and on ion species as well as the ion energy. Sputter yield and secondary coefficient data can be found in Chapman.[12] These metal atoms then travel through the system until they deposit on the substrate or other surface.

Because the electrons are trapped in the magnetic field, the plasma created is highly nonuniform. Charged particles traveling in a magnetic field are subject to the magnetic component of the Lorentz force, $\vec{F} = q(\vec{E} + \vec{v} \times \vec{B})$, and follow approximately helical trajectories around the magnetic field lines. The radius of the helix is called the Larmor radius or the cyclotron radius, and, for a singly charged particle, is given by $r_L = mv_\perp/eB$, where m is the particle mass, $v_\perp$ is the magnitude of the particle's velocity in the plane perpendicular to the magnetic field, e is the electron charge, and B is the magnitude of the magnetic field. In a typical magnetron sputtering system, with a magnetic field magnitude of several hundred Gauss, the Larmor radius for an electron in the region of strong magnetic field is much less than a millimeter, whereas that of an argon ion is several centimeters. As a result, the electrons, even those emitted from the target and entering the plasma with an energy of several hundred electron volts, are constrained by the magnetic field until they undergo a collision. Ions, on the other hand, due to their large Larmor radius, can readily reach the cathode even when their helical motion is centered on a magnetic field line several centimeters from the cathode. The plasma glow forms a loop adjacent to the target. A high-density plasma, with plasma density 10^{12} cm^{-3} or higher, exists in the loop, whereas the magnetron produces much lower plasma densities elsewhere.

In the circular planar magnetron configuration shown in Fig. 1, an overhead view would show the magnetic field lines as radial spokes forming a circle. The glow from the plasma in this configuration is doughnut shaped and located at the same radius, as is the etch track in the target. The fact that the field guides the plasma into a closed loop, circular or otherwise, is another essential feature of the magnetron. Unlike the electron shown in Fig. 2 electrons emitted from the cathode near the point at which the magnetic field is tangential to the target travel around the closed loop. As shown in Fig. 3, such an electron undergoes cycloidal motion. The electron's circular motion in the magnetic field is interrupted by the presence of the target, and the electron is reflected electrostatically by the sheath. The cycloidal motion will be repeated until the electron suffers a collision. Since all such electrons orbit the magnetic field lines in a right-hand sense, the cycloidal motion gives rise to an induced drift current around the loop.[13] If the loop were interrupted, electrons would drift out the end and magnetic confinement would be lost.

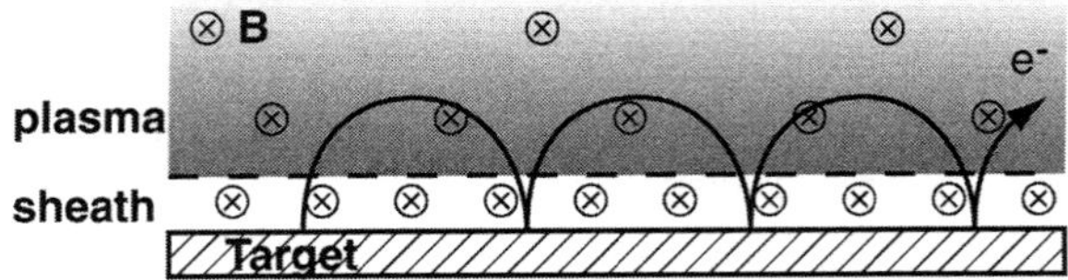

FIG. 3. Side view of planar magnetron, showing a surface on which $\vec{B}$ is parallel to the target face. Electrons emitted from the target at this point undergo cycloidal motion. The electron trajectory in the sheath is circular, and when electrons reenter the sheath they are reflected electrostatically.

When an electron undergoes cycloidal motion as shown in Fig. 3, it spends a greater fraction of its lifetime in the sheath, increasing the probability of ionization taking place in the sheath. Ionization in the sheath further increases the efficiency of the discharge, as the electrons born there may also pick up enough energy as they pass through the sheath to ionize more atoms. For the case shown in Fig. 3, the Larmor radius is larger than the sheath thickness. However, this is not always the case because the Larmor radius and the sheath thickness, s, depend differently on operating conditions. The Larmor radius, r_L, decreases with increasing magnetic field strength, and because the electron velocity increases with increasing cathode voltage amplitude, V_0, so does the r_L. When r_L is large compared to the sheath thickness s, s can be reasonably approximated by Child's law:

$$s = \frac{2}{3}\varepsilon_0^{1/2}\left(\frac{2e}{M}\right)^{1/4}\frac{V_0^{3/4}}{J_0^{1/2}}, \tag{1}$$

where J_0 is the current density, and M is the ion mass.[14] For typical operating conditions, the Child's law sheath thickness is on the order of millimeters and is thus comparable to r_L. The magnetron voltage varies relatively little as power is varied so that most of the variation in sheath thickness is derived from changes in the current density, which is a function of plasma density. As the plasma density increases with increasing magnetron power, the sheath thickness decreases.[15] When ionization in the sheath is substantial, Child's law is no longer valid and the expression for the sheath thickness must be modified.

The sputter rate depends on both the flux of ions to the target and the sputter yield per ion, which depends on the identity and the energy of the impinging ion as well as the angle of incidence. Argon is typically used as the working gas for metal sputtering. Because it is a noble gas, it does not react chemically with the metal. Also, because of its higher mass, it generally has a much higher sputter yield for a given energy than other inexpensive noble gases, such as helium or neon. The instantaneous sputtering rate is typically nonuniform across the sputter target. The sputter rate is maximum where the ion flux is highest, and this corresponds to the loop in the plane of the target face where the magnetic field is strong and tangential to the target surface. The system shown in Fig. 1 is a "sputter gun," in which the magnetic field configuration is axisymmetric so that the sputter track forms in a circle on the target. This is not always desirable from the standpoint of both target utilization and uniformity of deposition. Some commercial tools employ an asymmetric magnetic loop which is rotated slowly to generate a discharge which averaged over time is much more spatially uniform than when measured instantaneously.

Low-pressure operation (down to 1 mTorr) of the magnetron is desirable when the goal is to simply maximize deposition rate. At reduced pressure, scattering of sputtered metal atoms by collisions with the background gas is minimized. Since the metal atoms emitted from the target are directed preferentially toward the substrate, collisions with background gas atoms tend to deflect the metal atoms off course so they are more likely to deposit either back onto the target or on the chamber side walls. Thus, the higher the pressure, the lower the deposition rate at the substrate for a given sputter power. The electron mean free path for ionization increases with decreasing pressure, but as long as it is short compared to the system dimensions, the magnetron discharge operates normally. However, for very low pressures, it becomes difficult to sustain a magnetron discharge.

B. IMPLEMENTATION IN I-PVD PROCESS TOOLS

Magnetron sputter sources are used as metal vapor sources for I-PVD applications, in combination with some means to ionize the sputtered metal atoms. In order to achieve a high degree of ionization of the sputtered metal, I-PVD systems are generally operated at pressures higher than those used for conventional sputter deposition. Because the sputtered metal atoms are ejected with a substantial amount of energy, they traverse the system fairly rapidly at lower pressures. As a result, their residence time in the gas phase is relatively low, and so, therefore, is their probability of ionization. At higher pressures, collisions between the metal atoms and the background gas lead to a thermalization of the two populations; the metal is cooled while the gas is heated. Thermalization is desirable for I-PVD because it results in an increased residence time for the sputtered metal, increasing the probability of ionization. In general, the ionized metal flux fraction is higher at higher pressures. However, as mentioned previously, deposition rate

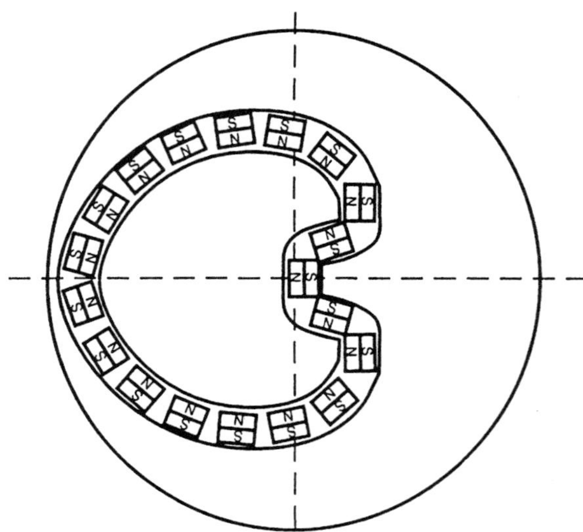

FIG. 4. Greater uniformity in target erosion is achieved through rotation of a puck containing a heart-shaped arrangement of permanent magnets. The puck is rotated about the center of the circular target, and the high-density magnetron plasma follows the magnetic fields as it rotates.[16]

declines with increasing pressure so that the choice of operating pressure is a trade-off between deposition rate and ionized metal flux fraction.

Because uniformity of the deposited film across the wafer diameter is very important in I-PVD (and in conventional magnetron sputtering), the arrangement of permanent magnets is of critical importance. The sputter rate on the target is highest where the magnetic field is tangential to its surface; therefore, a stationary magnet configuration produces a highly nonuniform erosion pattern on the target, making a uniform deposition rate very difficult. Translational motion of a "puck" assembly, shown in Fig. 4, containing the permanent magnets across the back of the target is an approach that has been taken to achieve much more uniform target erosion on average, even though the instantaneous erosion rate is nonuniform.[16]

III. Inductively Coupled Plasmas

ICPs are well-known for their use in plasma etch processes for ULSI fabrication,[17,18] but there are some important differences in their implementation for I-PVD processing. In both cases, radio frequency (RF) current through an antenna creates an electromagnetic field that couples energy to the plasma to sustain it. The mechanism is very similar to induction heating of conducting materials, in which in this case the plasma is the conducting material. Electrons gain energy through acceleration in the RF electric field and, in turn, the energetic electrons ionize gas atoms or molecules, replenishing the plasma that is lost to the walls through diffusion. In typical configurations for ICP plasma etch tools, the RF antenna is located just outside the vacuum chamber. To allow the RF electromagnetic fields to penetrate into the vacuum chamber and sustain the plasma, a dielectric material (typically quartz) is required for the vacuum wall separating the antenna from the plasma. The RF fields will not pass through a continuous metal wall; an induction antenna placed next to a metal surface will instead result in induction heating of the metal wall. For this reason, it is not practical to use a conventional ICP source with such an external antenna for I-PVD processing. The problem is that although a dielectric vacuum window might be installed in an I-PVD system, it will quickly become coated with a layer of sputtered metal which prevents inductive coupling of power to the plasma.

In this section, we follow the evolution in ICP tool design for plasma processing during the past several years. I start with a discussion of ICP tool design as developed for plasma etching applications. Following this is a description of variations on that design developed to address the problem of metal deposition on dielectric surfaces.

A. Principle of Operation

In ICPs for plasma etching, the discharge is sustained through inductive coupling of RF power (typically 13.56 MHz, but also as low as 450 kHz) to the plasma from an antenna external to the vacuum chamber. The antenna typically has a helical or flat spiral geometry and is separated from the plasma by a quartz vacuum wall. Capacitive coupling of power to the plasma, due to high voltages associated with the antenna currents, may also be significant. ICP discharges are one of several types of "high-density" discharges, along with ECR and helicon, that offer high plasma densities (10^{11}–10^{12} cm^{-3}) and low operating pressures (1–10 mTorr) compared to parallel plate, capacitively coupled discharges used for semiconductor processing since the 1970s. Besides the capability of high plasma densities at low pressure operating conditions, another desirable feature of ICP tools used in plasma etching is that they are "electrodeless." In parallel plate process tools, the RF-powered electrode experiences voltages sufficient to sputter electrode material, which poses the threat of contamination in etching processes for sensitive electronic devices. With the antenna outside of the chamber, this is not an issue with ICP plasmas. ICP discharges for plasma etching processes have been experimentally characterized and modeled extensively[19–28] and have been described in several review articles and book chapters.[14,29,30]

A planar RF ICP discharge similar to those used in etching tools is illustrated in Fig. 5. RF currents running through the spiral antenna comprise what can be thought of as a sequence of concentric magnetic dipoles, with fields of neighboring turns interfering constructively in the plasma. Approximate magnetic field lines in the presence of a plasma are also depicted in Fig. 5. An azimuthal RF electric field is induced by the time-varying magnetic field and is proportional to ωB, where $\omega = 2\pi f$ is the RF frequency. The RF electric field provides direct heating of electrons which comprises the power input to the plasma. An image current is induced in the plasma in the direction opposite the coil current so that field penetration into the plasma is limited to an axial distance characterized by the skin depth, δ, which is typically on the order of 1 or 2 cm.[26] In contrast to ECR and helicon discharges, there is no propagating wave in the plasma, and power coupling is strictly a surface phenomenon.

Inductive RF power deposition can be approximately divided into highly collisional and collisionless regimes, with qualitatively different behavior depending on whether electron collisions with neutrals are numerous on the RF time scale.[31] This is determined by comparing the electron-neutral collision frequency, ν_{en}, with the RF frequency, ω. In the collisional case, $\nu_{en} \gg \omega$, ohmic heating occurs as power is absorbed through electron collisions with neutrals. However, most commercial ICP etching sources operate over a range of frequencies and neutral pressures where $\nu_{en}/\omega \lesssim 1$. In the collisionless regime, electrons typically pass through the region of

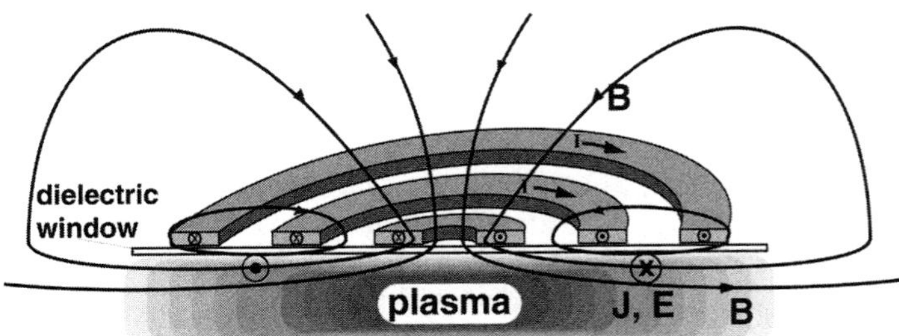

FIG. 5. Schematic diagram of a planar RF inductively coupled plasma. This design, developed for plasma etching applications, is not practical for I-PVD applications because coating of the dielectric window by a conducting film will interfere with RF power coupling to the plasma.

strong electric field without undergoing collisions with neutrals, dissipating the energy they gained elsewhere.[31] Since I-PVD tools run at somewhat higher pressures, some collisional, or ohmic, electron heating is likely.

B. ICP DESIGN FOR I-PVD

Because of the problem of metal coating of the dielectric window, inductively coupled plasma tools designed for etching processes are not appropriate for I-PVD. Two alternative approaches are described here. The first design has an induction antenna inside the vacuum system so no window is required.[32,33] The second has an external antenna and dielectric window but also has an internal slotted Faraday shield structure that prevents coating of the dielectric window.[34]

The metal introduced through sputtering influences the properties of the high-density ICP discharge. Although the partial pressure of the metal vapor may be significantly lower than that of argon, it may have a major impact for several reasons. First, the ionization energy of the metal atoms may be significantly lower than that of argon. For example, copper has an ionization energy of 7.72 eV, compared to 15.76 eV for argon. As a result, the metal atoms are "easier" to ionize in collisions with electrons, so a substantial fraction of the ions may be metal.[35]

One might expect that, due to the lower ionization energy of metal atoms, the plasma density would increase with the addition of metal for a given RF power to the antenna. However, exactly the opposite is the case,[36] and the reason is ascribed to gas rarefaction by the sputtered metal.[36–38] Sputtered metal atoms are released from the surface with an energy on the order of several electron volts, and the argon background is heated through collisions with the energetic metal atoms.[9] The resulting gas density leads to lower plasma densities for a given RF power level. ICP discharges (as well as most other discharge types) are less efficient at channeling electron energy into ionization at lower pressures[20] due to enhanced energy losses at the chamber walls.

1. Internal Antenna Configuration

In the internal antenna configuration, a helical antenna of one or more turns is located inside the vacuum chamber,[39] with a radius larger than the radius of the substrate. A typical arrangement is shown in Fig. 6. Electrical connections to the two ends of the antenna are made through vacuum feedthroughs, and a baffle must be used to prevent shorting by the deposited metal.

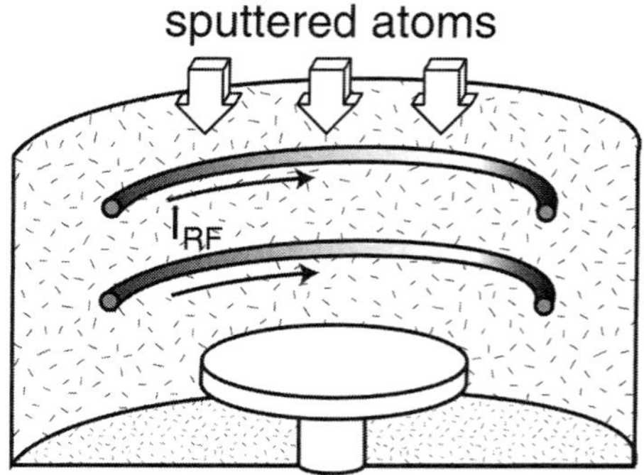

FIG. 6. Internal induction coils for the I-PVD process.

The plasma is generated in much the same way as that for the planar ICP with external antenna. Azimuthal electric fields, localized in the vicinity of the antenna, are induced by RF currents in the antenna. The RF electric fields couple energy into electrons which in turn ionize the gas. Although electron heating is localized near the outer radius of the chamber, diffusion of the plasma toward the chamber center may result in a plasma density profile that is peaked on axis.[34] Diffusion profiles are affected by chamber geometry and gas pressure.[27] If the chamber is too short or the pressure too high, the density maximum could move out radially toward the antenna, impacting uniformity. This is unlikely for typical current system designs, but it could be a factor in tools for 300-mm wafers.

A greater degree of capacitive coupling of RF power can be expected in a system with an internal induction antenna than in one with an external antenna, which could lead to somewhat lower plasma densities. While inductive coupling of RF power results from the electromagnetic fields induced by current flow through the antenna, capacitive coupling arises from the fields associated with the antenna voltage. In a parallel plate RF discharge, power coupling to the plasma is entirely capacitive. In capacitive coupling, electrons are heated through acceleration in the electric fields between the electrodes. Parallel plate discharges are generally less efficient than inductive discharges because much of this field is concentrated in sheaths, in which a substantial fraction of the power goes into acceleration of ions across the sheath and into the electrode. Although some degree of capacitive coupling also occurs in inductive discharges, ion heating may be a much smaller sink of RF power in inductive discharges. In an ICP discharge with an external antenna, voltages of amplitude 1 kV or larger may develop across the antenna; however, most of the voltage drop occurs across the dielectric window so that the voltages seen by the plasma are relatively modest. One can think of the combination of the dielectric window and the sheath as two capacitors in series, forming a capacitive voltage divider. Due to its substantial thickness, the dielectric window has a higher capacitive impedance[14] and carries most of the voltage drop. In a system with an internal induction antenna there is no dielectric between the antenna and the plasma, so the voltage drop is all across the sheath and substantial capacitive coupling is possible. On the other hand, the lack of a dielectric layer between the antenna and the plasma improves inductive coupling so that the antenna voltages are not as high as those for an external antenna delivering a comparable amount of RF power.

Sputtering of the antenna is an important consideration when the antenna is inside the vacuum system.[40,41] The sputtered antenna material will deposit on the substrate along with the material from the sputter gun, potentially contaminating the film. Sputtering of the antenna occurs because a negative DC self-bias voltage develops on the antenna, much as it does on a RF-powered substrate holder. The self-bias arises from the mass difference of electrons and ions. Internal ICP antennas may develop RF voltage amplitudes of hundreds of volts. When the voltage swings positive with respect to the plasma, electrons are collected; when the voltage swings negative, ions are collected. The thermal velocity for electrons is $\sqrt{8\,kT_e/\pi m_e}$, where k is the Boltzmann constant and T_e is the electron temperature. Ions collected at a surface will enter the sheath with the Bohm velocity, $\sqrt{kT_e/M_i}$. Thus, because of the lighter mass of the electrons, their flux to the surface is higher than that of the ions and net negative charge is collected. As the negative charge accumulates, the voltage (averaged over the RF period) decreases until a steady state is reached in which most of the electrons are repelled and the net fluxes of ions and electrons are equal.[12]

2. *Shielded Antenna Configuration*

Faraday shields are generally used with RF antennas for induction heating to block capacitive coupling of RF power between the antenna and the plasma while allowing induction fields to pass. However, it has another very important function when applied to an I-PVD system, namely, it prevents coating of the dielectric window by sputtered metal.

For example, a Faraday shield and a spiral antenna are shown in Fig. 7. The spoked Faraday shield, cut from sheet metal, is located between the antenna and the plasma. The shield may be located either on the atmosphere or on the vacuum side of the dielectric window (although it is unlikely to be on the atmosphere side for I-PVD), but it must be electrically isolated from the antenna. The key feature of the shield is the slots between the spokes. If it were a solid disk, image currents in an approximately

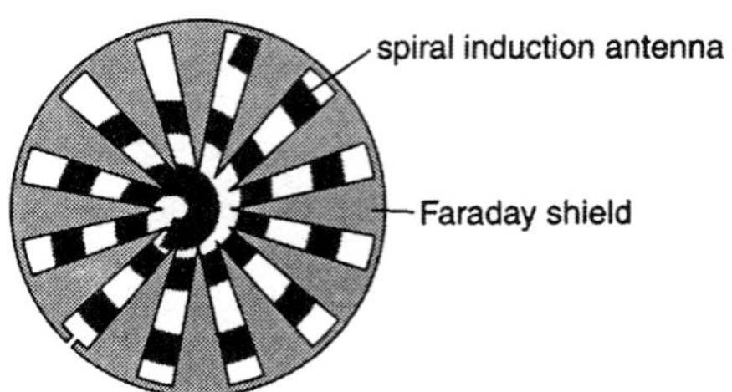

FIG. 7. A Faraday shield is shown with a spiral induction antenna. The key element of the Faraday shield is the slots.

circular pattern would be induced in the shield when the antenna was driven by RF currents. Passage to the plasma of RF fields would be blocked, and whatever wave energy did not go into induction heating of the disk would be reflected. However, slots cut into the Faraday shield (both between the spokes and in the rim) perpendicular to the direction of current flow break up the circular current path and allow radiation to pass into the plasma.

In the design developed for I-PVD by Dickson *et al.*,[34] a helical antenna around a dielectric tube is used, along with an internal Faraday shield shown in Fig. 8. The Faraday shield has a slightly smaller diameter than the inside of the quartz tube so that there is a small gap, and it requires only a single vertical slot. The dielectric window will be coated along a thin vertical strip exposed by the slot in the Faraday shield. However, because of the gap, the shield is not in physical contact with the dielectric wall so that there is no electrical connection between the shield and deposited metal on the

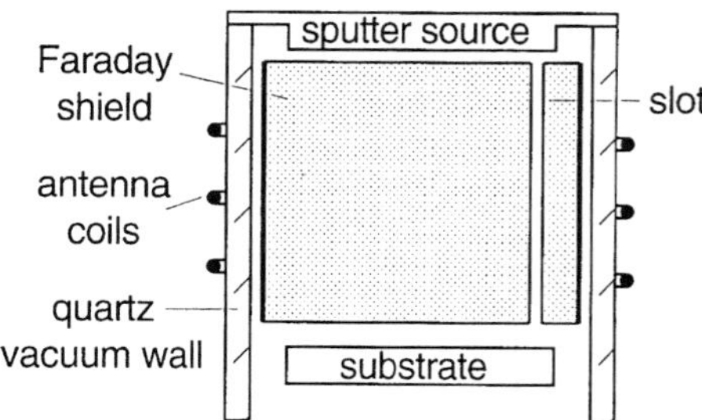

FIG. 8. Inductively coupled plasma system for ionization of sputtered metal. The antenna is wound around a quartz tube, and a Faraday shield is located inside the vacuum chamber. The vertical slot in the Faraday shield is necessary to allow induction fields to pass. A gap between the Faraday shield and quartz wall prevents shorting between the shield and metal deposited on the quartz wall.

dielectric. Therefore, there is no closed current path around the circumference, thus blocking the flow of induced RF currents on the inner wall. In the absence of the shield, a continuous film would form around the inner surface of the dielectric wall so currents could flow, preventing RF power flow to the plasma.

It is also worth noting that the source described by Dickson *et al.*[34] uses a DC sputter source *without* a magnetron magnetic field. In this case the inductively coupled plasma provides a sufficiently high-density plasma in the vicinity of the sputter target that the additional magnetic confinement is unnecessary to achieve acceptable sputter rates.

IV. Electron Cyclotron Resonance Plasmas

ECR discharges, adapted for plasma etching applications in the late 1970s,[42] have been employed for I-PVD processes as high-density plasma sources to ionize metal atoms introduced with either a DC sputter source or some other source of metal.[43–48] Two features of ECR discharges make them an attractive choice for a high-density discharge. First, energy is efficiently coupled to the electrons responsible for maintaining the discharge. Second, the DC magnetic field used in ECR discharges provides plasma confinement. The magnetic confinement enhances plasma density while permitting lower pressure operation. Like ICP discharges, ECR plasmas are electrodeless, efficiently producing high plasma densities while maintaining low sheath voltages. However, slightly higher plasma densities are possible with ECR discharges (densities close to $10^{12}\ cm^{-3}$ have been reported for I-PVD plasmas[49] because microwaves in an ECR discharge propagate through the plasma to heat electrons in the middle of the discharge, whereas electron heating in an ICP discharge takes place only near the antenna surface. ECR sources have been implemented for a variety of applications, including plasma etching for microelectronics fabrication and I-PVD, and are described in detail in a recent review article.[50]

A. PRINCIPLE OF OPERATION

An ECR discharge takes advantage of a DC magnetic field and a microwave power source to produce a high-density plasma. For any ECR system, there is an ECR zone, a region in which there exists a DC magnetic field strength for which the frequency of electron cyclotron motion matches the frequency of microwave excitation. Many configurations of DC magnetic field and microwave applicators have been developed.[50] A typical configuration,

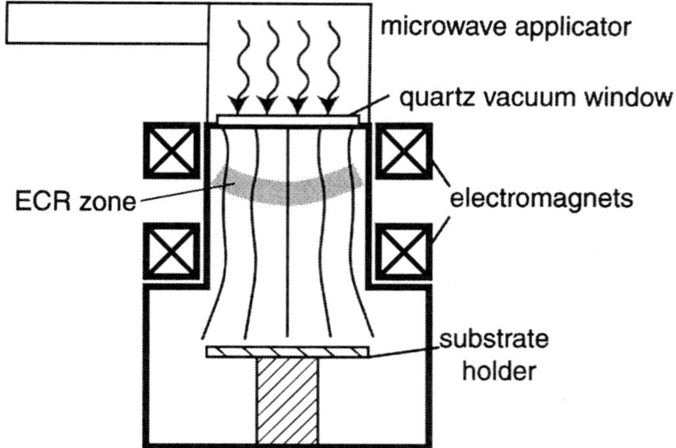

FIG. 9. Example of ECR system design for plasma processing applications. This design is unsuitable for I-PVD applications because coating of the quartz vacuum window with a conducting material would block microwaves from entering the discharge.

shown in Fig. 9, consists of a waveguide introducing microwaves traveling parallel to an approximately axial DC magnetic field through a quartz vacuum window at the top of the vacuum system.

As discussed in the section on DC magnetron plasmas, electrons in a DC magnetic field experience approximately helical trajectories, and the same holds in an ECR discharge. The frequency at which electrons orbit around the magnetic field lines, $\omega_{ce} = eB/m$, the cyclotron frequency, depends on the local magnetic field strength but is independent of electron energy. In some region of any ECR discharge, the cyclotron frequency of the electrons must match the frequency of the electromagnetic wave providing the energy to sustain the plasma. In this region, the wave energy is strongly absorbed by the discharge electrons. A common choice of microwave frequency for ECR discharges is 2.45 GHz, and the corresponding magnetic field strength for cyclotron resonance is 875 G.

Resonant energy absorption by electrons occurs when the direction of the electron helical motion remains aligned with the electric field of the electromagnetic wave throughout the oscillation cycle.[14] The resonant electron is thus continually accelerated in the wave electric field, and the resonant absorption of energy damps the wave. As the electron velocity increases, the radius of its orbit, the cyclotron radius, also increases.

B. MODIFICATION OF CONVENTIONAL DESIGN FOR I-PVD

There are several challenges associated with adapting an ECR system for I-PVD processes. The geometrical configuration of the system must accommodate the introduction of both the microwaves and the metal vapor. This is achieved in ICP systems by placing the sputter source on axis at the top of the chamber, and RF power is introduced at the outer radius through helical antennas. Another approach taken with ECR is to use an annular sputter target at the top of the chamber and introduce microwaves on axis through the hole in the target, as shown in Fig. 10.[49] In this design, no magnetron magnet is required at the sputter target for efficient sputtering. The ECR source provides a sufficiently high-density plasma near the target so that additional magnetic confinement is not required. A negative bias voltage of several hundred volts is applied to the sputter target to accelerate ions from the ECR plasma into the target.

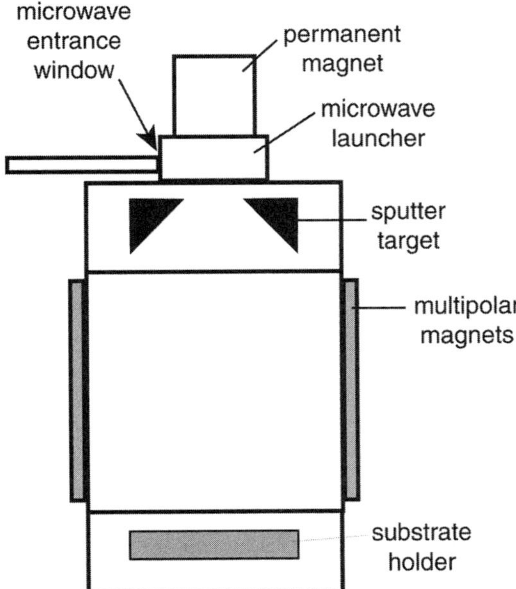

FIG. 10. ECR system design for I-PVD. Microwaves enter the chamber through a hole in the annular sputter target. The vacuum window is recessed to prevent coating with a metal film. Multipolar magnets (alternating strips of north and south poles) provide additional confinement for downstream plasma.[38,49]

1. Remote Vacuum Window in Waveguide

As in inductively coupled plasmas, the power to sustain an ECR discharge in typical configurations is coupled through a dielectric window. This presents a problem in an I-PVD system in which windows exposed to the discharge will be coated with a thin metal film. The waves will be reflected by the metal film rather than passing through it and will not make it to the plasma.

A remote vacuum window has been successfully implemented to avoid coating of the dielectric vacuum window in the microwave applicator with metal, which would block the microwaves from entering the discharge.[43,49] Rather than the common location at the entrance to the plasma source, the window is recessed far enough back in the waveguide so that it is not in the plasma's line of sight. An example is shown in Fig. 10.

V. Hollow Cathode Magnetron Discharges

A. SPUTTERING AND IONIZATION IN A SINGLE PLASMA SOURCE

Hollow cathode magnetron discharges have been introduced as a high-density plasma source for I-PVD processes[51] in addition to earlier introduction as ion sources.[52–54] Hollow cathode magnetrons, a magnetized version of the hollow cathode, offer both sputtering and ionization of the sputtered material in a single source, an advantage over other methods. Furthermore, the sputtered material leaving the hollow cathode source may have a higher degree of ionization than that for the other sources, leading to improved characteristics of the deposited I-PVD films. Although their use as I-PVD sources is relatively recent, hollow cathode discharges have been studied for some time, and the concept has been developed for applications including spectroscopic light sources, gas lasers, and ion sources. Due to specific features of the electrode geometry, current densities in a hollow cathode discharge may greatly exceed those observed in a planar discharge at the same voltage. The electrode geometry in hollow cathode discharge enables more efficient use of both energetic electrons and ions in maintaining the plasma. In hollow cathode discharges, energetic electrons are electrostatically confined and can oscillate between opposed cathode surfaces (the "pendulum" effect) so that most of the energy acquired in the cathode fall is used in sustaining the plasma.

Many types of discharges, such as magnetrons discussed previously, are sustained, at least in part, by γ electrons, secondary electrons emitted from the cathode as a result of ion bombardment. The γ electrons pick up energy from the sheath electric field and expend part of it in ionization to replenish plasma lost through diffusion to the walls or through volume recombination. The rest of the electron energy is divided between excitation of atoms or molecules, molecular dissociation, and being transferred to the walls in the form of heat when the electrons hit the wall. The efficiency with which electron energy is expended in ionization compared to these other pathways is a major factor in determining plasma density or whether a plasma can be sustained at all for a given set of operating conditions. The efficiency of these so-called γ discharges also depends on the fraction of ions that reach the cathode surface as they escape from the discharge since the flux of γ electrons into the discharge is directly proportional to the flux of ions to the cathode.

Several enhancements to a simple parallel plate configuration have been devised to improve ionization efficiency, including harnessing a greater proportion of discharge power for ionization compared to other processes, such as losses to the walls. These enhancements enable higher plasma densities and operation at pressures lower than would otherwise be possible. One example of such an enhancement is the magnetron discharge described previously. Another example is the hollow cathode discharge, in which enhanced efficiency is achieved by appropriate choice of the mechanical geometry of the cathode. Hollow cathode discharges have been developed for ion sources,[55] spacecraft charge control, plasma sources for diagnostic development,[56] electron gun,[57] as well as other applications. Although there are many variations, the distinguishing feature of hollow cathode configuration is a concave cathode structure with opposing cathode surfaces between which electrons are electrostatically confined. The hollow cathode configuration not only leads to higher plasma densities for a given voltage but also enables an increased fraction of ions to reach the cathode/sputter target, increasing the sputter efficiency. Because of the geometry of the hollow cathode configuration, sputtered atoms reaching the substrate have a higher probability of being ionized than they do with other types of discharges. As a result, the hollow cathode can function as a source of metal ions, without the use of a secondary plasma as required in other I-PVD approaches. The "conventional" (no added magnetic field) hollow cathode is not in common use as a sputter source, but sputtering has been reported as a by-product of operation.[58]

The hollow cathode magnetron combines features of the conventional hollow cathode with the magnetron described in Section II. Introduced in 1996 by Helmer and coworkers[51] for I-PVD, the hollow cathode magnetron features a magnetron magnetic field configuration inside the hollow cathode. Although introduced relatively recently, some of the elements can be seen in earlier studies. Yeom *et al.*[59] describe cylindrical magnetron discharges, with a cylindrical cathode and an axial magnetic field. Window *et al.*[60] achieved a high ion flux to the substrate with a pair of facing magnetron cathodes.[60] Cuomo and Rossnagel[61] introduced a small hollow cathode tube to augment a planar magnetron discharge and enhance its performance.

B. PRINCIPLE OF OPERATION

1. "Conventional" Hollow Cathode

In order to describe the hollow cathode effect in more detail, a brief review of a planar DC discharge, as shown in Fig. 11, may be helpful. The anode is grounded, and the cathode is biased to a negative potential of several hundred volts, resulting in a potential profile between the electrodes, $V(x)$, similar to that shown in Fig. 11. In a sputtering system, the cathode is also referred to as the target. A plasma consisting of electrons and positive ions (no negative ions) tends to acquire a potential slightly higher than that of the surfaces exposed to it. This sets up a sheath electric field at all plasma/surface boundaries directed from the plasma toward the wall. This sheath electric field serves to confine electrons, which, due to their lower mass and thus higher mobility, otherwise have a tendency to diffuse to the walls at a higher rate than the ions.

Ions reaching the cathode are accelerated in the sheath electric field and strike the cathode surface with an energy of hundreds of electron volts. This energetic ion bombardment of the cathode surface results not only in sputtering of the target material but also in secondary emission of electrons. The secondary emission coefficient, γ, is defined as the average number of electrons emitted per ion striking the surface. The value of γ depends in

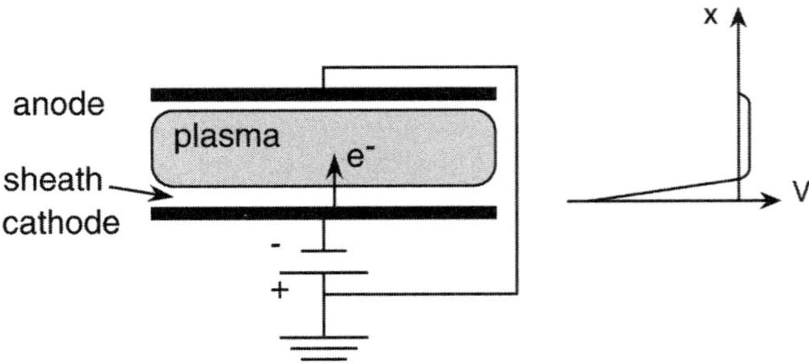

FIG. 11. Parallel plate DC discharge, showing electrostatic potential profile between electrodes.

general on the energy of the ion striking the surface, but it is generally less than 1. The emitted electrons are accelerated in the opposite direction through the sheath back into the plasma, also picking up energies of several hundred electron volts. Some of the energy of these γ electrons is expended in ionization, replenishing the ions that are lost through diffusion to the cathode or other surfaces. However, the γ electrons are generally directed toward the anode, and some will reach the anode surface, easily overcoming the potential barrier at the anode sheath and transferring their energy to the wall. Sustaining a discharge depends on a sufficient fraction of the energy of the γ electrons contributing to ionization, so the energy carried by γ electrons to the anode is wasted. Electron energy loss at the anode ultimately places a limitation on the ability to maintain a discharge at low pressures where the mean free path of the energetic electrons becomes comparable to the distance between cathode and anode.

The primary difference between conventional hollow cathode discharges and planar DC discharges is the concave shape of the cathode, as illustrated in Fig. 12.[62–64] Energetic γ electrons produced through ion bombardment secondary electron emission from the cathode still have primary responsibility for maintaining the discharge, but they are now electrostatically trapped inside the cathode and oscillate between opposite walls of the cathode interior (the pendulum effect).[65] Several factors associated with the hollow cathode effect (HCE) combine to produce much higher current densities for a given voltage than a planar discharge. Due to the trapping, a greater fraction of the γ electron energy is expended in ionization before loss to the anode, resulting in higher density plasmas and allowing lower pressure operation.

Although the pendulum effect clearly cuts down on losses of fast electrons to the anode, this is not the only factor that makes hollow cathode discharges more efficient than planar DC discharges. Another consequence of the pendulum effect is the increased possibility of ionization by fast electrons in the cathode sheath. Each time a fast electron traverses the

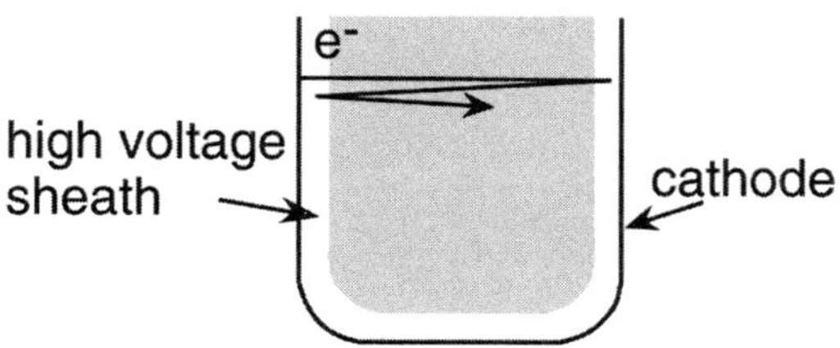

FIG. 12. In hollow cathode discharges, energetic γ electrons are trapped electrostatically in the concave cathode structure, enhancing ionization efficiency.

hollow cathode, it enters the sheath as it turns around. Progeny electrons created in the sheath are then accelerated to high energy by the strong sheath electric field, and they too contribute to ionization.

Greater containment of ions in the vicinity of the cathode due to the geometrical configuration of the hollow cathode is also a factor in the HCE. The fraction of ions lost to the anode is thus lower than that for a planar discharge. The resulting higher flux of ions to the cathode leads to increased production of γ electrons and a greater sputter rate.

2. *Hollow Cathode Magnetron*

The hollow cathode *magnetron* developed for I-PVD applications is a DC hollow cathode with the added feature of a DC magnetic field.[51] The hollow cathode has an opening width of several centimeters and is surrounded by permanent magnets (Fig. 13). The hollow cathode magnetron may be cylindrically symmetric, but other configurations, such as a ring shape and others, have been suggested to improve spatial uniformity. In either case, the

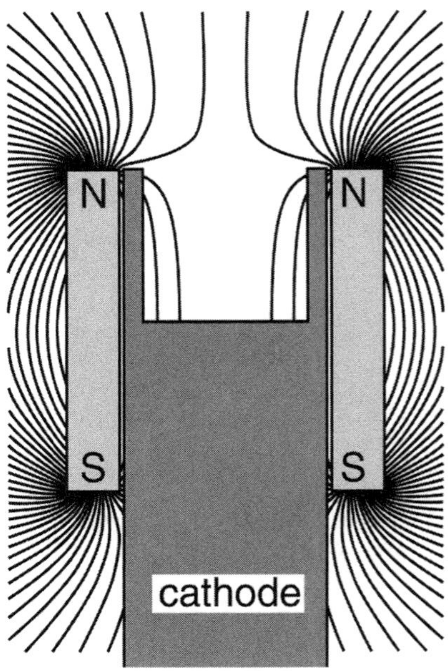

FIG. 13. A cylindrical hollow cathode magnetron consists of a hollow cathode surrounded by an annular permanent magnet, with magnetic field lines that pass through the interior of the hollow cathode.

cross section of the hollow cathode would appear approximately as depicted in Fig. 13. A magnetron magnetic field configuration is formed at the inner walls of the cathode. Some magnetic field lines intersect the cathode surface twice, as in the case of the DC planar magnetron. Gamma electrons are confined in a doughnut-shaped region defined by these magnetic field lines. In fact, due to the cylindrical magnetron formed in the interior of the hollow cathode, the electrostatic confinement of γ electrons described in the HCE may not be important in the hollow cathode magnetron.

Ion containment near the cathode, on the other hand, is an aspect of the HCE that is critical to the hollow cathode magnetron and is enhanced by the presence of the magnetic field. Figure 13 shows magnetic field lines for a cylindrically symmetric hollow cathode magnetron. In the center of the hollow cathode is a region of relatively weak magnetic field in which a high-density plasma forms. In addition, there is a cusp magnetic field in the vicinity of the opening of the hollow cathode that provides magnetic confinement for the high-density plasma. To define a magnetic cusp, one might think of a pair of permanent bar magnets oriented on a line and with the two north poles close together and facing one another. At the magnet surface the field lines are directed toward the other magnet, but then they must bend sharply to terminate at the south pole rather than the nearby north pole. In this case the cusp arises from the fields emanating from the two north poles visible on opposite sides of the hollow cathode in Fig. 13. A null point in the magnetic field occurs on the axis near the opening of the hollow cathode, with the field directed upward above the null point and downward below the null point. The plasma that escapes the cusp is guided along the field lines extending axially from the hollow cathode. The high-density plasma inside the hollow cathode is critical to achieving a high degree of ionization of the sputtered metal.

The confinement of the plasma in the cusp magnetic field can be explained in terms of single particle motion. As mentioned previously, an electron in a magnetic field moves in an approximately helical trajectory, with the axis of the helix parallel to the direction of the local magnetic field. If the electron moves into a region of stronger magnetic field (i.e., field lines closer together), its velocity along the magnetic field line decreases while the perpendicular velocity increases. An electron moving in any direction from the null point will be heading into a region of stronger magnetic field. Eventually, motion along the field line may stop so that the electron reverses its direction along the field line and is "reflected" back into the region from which it came. This is known as the mirror effect and is described in detail in the literature.[66] Thus, lower energy electrons tend to be confined in the cusp region, whereas those with sufficient velocity parallel to the magnetic field may make it through.

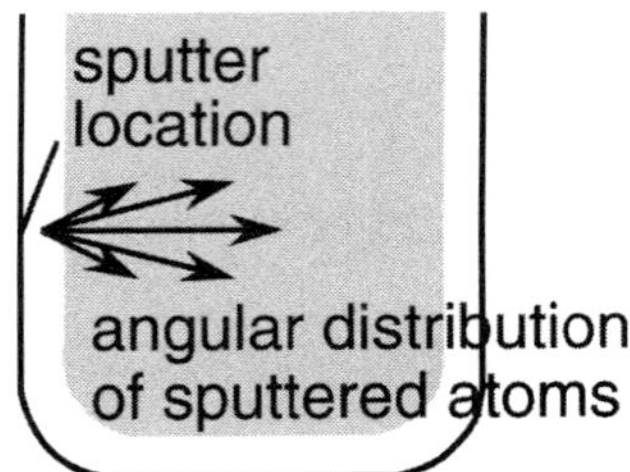

FIG. 14. Atoms sputtered from the hollow cathode sidewalls are preferentially directed toward the opposite wall. A large fraction of the metal that exits the hollow cathode and reaches the substrate is in ionized form.

The implication for the hollow cathode magnetron is that a high-density plasma (with plasma densities reported as high as $10^{13}\ \text{cm}^{-3}$) exists in the interior of the hollow cathode. This high-density plasma is made up of two distinct components, the magnetron plasma that hugs the inner cylindrical wall of the cathode and the plasma that is magnetically trapped near the cathode opening. Therefore, the metal atoms sputtered from the hollow cathode come in contact with a high-density plasma, increasing the likelihood that they too will be ionized.

An additional feature of the hollow cathode magnetron that enhances the ionization fraction of the metal reaching the substrate is illustrated in Fig. 14. Material sputtered from surfaces is generally directed perpendicular to these surfaces, with some spread in the velocity distribution as indicated in Fig. 14. Therefore, metal sputtered from the magnetron region on the inner wall of the hollow cathode is preferentially directed toward the opposite wall. This lessens the probability that a neutral sputtered atom will reach the substrate. Once ionized in the high-density plasma, however, the metal ion may be guided toward the substrate by electrostatic fields in the discharge.

VI. Comparison of High-Density Plasma Sources

The sources described in this chapter can be divided into two categories: those for sputtering metal and those for ionizing the sputtered metal, although the lines between them sometimes become blurred. For both types, perhaps the biggest challenge is uniformity in the fluxes of both metal atoms and ions at the wafer surface.

The most widely used approach to sputtering metal has traditionally been the magnetron, but for I-PVD there are alternatives. First, when used in combination with another high-density plasma source, the magnetron magnetic field may be unnecessary. The purpose of the magnetic field is to *create* a high-density plasma adjacent to the target so it can be eliminated if there is another high-density plasma source, as demonstrated for both ICP and ECR systems.[34,49] However, although elimination of the magnetron magnets is desirable from the practical perspective of simplifying design and maintenance, it may expose wafers to damaging bombardment by energetic γ electrons. Second, the hollow cathode magnetron is a variation of the magnetron that produces a high degree of ionization of the sputtered metal.

ICP and ECR have both been examined as secondary sources for ionizing gas-phase metal atoms. Both require modification from standard plasma etching configurations to avoid deposition on dielectric windows. Higher plasma densities are possible with ECR than with ICP, but there is a price and this does not necessarily translate to a higher degree of ionization of the metal vapor.[38] The plasma density in an inductive source is self-limiting because the higher the density, the more effective the plasma is in shielding out the RF fields. In ECR an electromagnetic wave propagates through the plasma, and densities as high as $10^{13}\,\text{cm}^{-3}$ are possible. However, microwave power inputs of 2–5 kW are required to achieve densities this high, and low-pressure (0.1–1 mTorr) operation is necessary under these conditions to prevent arcing in the microwave launcher. At lower pressures the mean free path for ionization of sputtered metal atoms becomes longer, and a lower ionized metal flux fraction is realized at the substrate.[38] ICP sources have the advantage of simplicity compared to ECR sources. The DC magnetic field required for ECR sources adds expense and bulk to the tool design and creates problems for uniformity and scaling to larger substrate size.

The hollow cathode magnetron appears to be very effective at ionizing metal atoms in a localized region (the interior of the hollow cathode). The width of the hollow cathode is necessarily restricted to achieve this effect, so scaling the source to produce uniform deposition rate and film quality presents different challenges than those of the other options.

Another common type of high-density plasmas for materials processing are helicon sources. Helicon sources have only recently been adapted for I-PVD. In helicon sources an electromagnetic wave (typically RF) is launched with an antenna to propagate parallel to the axis of a cylindrical plasma, magnetized in the axial direction. Like ECR, because the energy for electron heating is carried by a wave propagating into the plasma, very high plasma densities are possible. An I-PVD system using a magnetron sputter source and a helicon plasma as a secondary source of high-density plasma has been reported.[67] One or more compact helicon sources may be mounted on the side of the chamber so that the plasma is introduced through sideports connecting the helicon to the main chamber.

This chapter has highlighted several of the most widely used high-density plasmas for plasma processing and particularly I-PVD, but it by no means provides an exhaustive list. Other approaches are available, and new ones will no doubt be developed.

References

1. J. A. Thornton, *J. Vac. Sci. Technol.* **15**, 71 (1978).
2. J. A. Thornton, *J. Vac. Sci. Technol.* **15**, 188 (1978).
3. J. C. Helmer and C. E. Wickersham, *J. Vac. Sci. Technol. A* **4**, 408 (1986).
4. S. M. Rossnagel and H. R. Kaufman, *J. Vac. Sci. Technol. A* **4**, 1822 (1986).
5. B. Window, F. Sharples, and N. Savvides, *J. Vac. Sci. Technol. A* **3**, 2368 (1985).
6. S. M. Rossnagel and H. R. Kaufman, *J. Vac. Sci. Technol. A* **5**, 2276 (1987).
7. A. E. Wendt, M. A. Lieberman, and H. Meuth, *J. Vac. Sci. Technol. A* **6**, 1827 (1988).
8. S. M. Rossnagel and H. R. Kaufman, *J. Vac. Sci. Technol. A* **6**, 223 (1988).
9. S. M. Rossnagel, *J. Vac. Sci. Technol. A* **6**, 19 (1988).
10. T. E. Sheridan, M. J. Goeckner, and J. Goree, *J. Vac. Sci. Technol. A* **8**, 1623 (1990).
11. A. E. Wendt and M. A. Lieberman, *J. Vac. Sci. Technol. A* **8**, 902 (1990).
12. B. Chapman, *Glow Discharge Processes*, Wiley, New York, 1980.
13. S. M. Rossnagel and H. R. Kaufman, *J. Vac. Sci. Technol. A* **5**, 88 (1987).
14. M. A. Lieberman and A. J. Lichtenberg, *Principles of Plasma Discharges and Materials Processing*, Wiley, New York, 1994).
15. L. Gu and M. A. Lieberman, *J. Vac. Sci. Technol. A* **6**, 2960 (1988).
16. R. E. Demaray, J. C. Helmer, R. L. Anderson, Y. H. Park, R. R. Cochran, and J. V. E. Hoffman, Rotating sputtering apparatus for selected erosion, United States Patent No. 5,252,194, 1993.
17. J. B. Carter, J. P. Holland, E. Peltzer, B. Richardson, E. Bogle, H. T. Nguyen, Y. Melaku, D. Gates, and M. Ben-Dor, *J. Vac. Sci. Technol. A* **11**, 1301 (1993).
18. R. Patrick, P. Shoenborn, H. Toda, and F. Bose, *J. Vac. Sci. Technol. A* **11**, 1296 (1993).
19. M. S. Barnes, J. C. Forster, and J. H. Keller, *Appl. Phys. Lett.* **62**, 2622 (1993).
20. L. J. Mahoney, A. E. Wendt, E. Barrios, C. J. Richards, and J. L. Shohet, *J. Appl. Phys.* **76**, 2041 (1994).
21. D. F. Beale, A. E. Wendt, and L. J. Mahoney, *J. Vac. Sci. Technol. A* **12**, 2775 (1994).
22. V. I. Kolobov, D. F. Beale, L. J. Mahoney, and A. E. Wendt, *Appl. Phys. Lett.* **65**, 537 (1994).
23. P. L. G. Ventzek, R. J. Hoekstra, and M. J. Kushner, *J. Vac. Sci. Technol. B* **12**, 461 (1994).
24. U. Kortshagen, I. Pukropski, and L. D. Tsendin, *Phys. Rev. E* **51**, 6063 (1995).
25. H. Sugai, K. Nakamura, Y. Hikosaka, and M. Nakamura, *J. Vac. Sci. Technol. A* **13**, 887 (1995).
26. J. A. Meyer and A. E. Wendt, *J. Appl. Phys.* **78**, 90 (1995).
27. J. A. Stittsworth and A. E. Wendt, *Plasma Sources Sci. Technol.* **5**, 429 (1996).
28. M. J. Kushner, W. Collison, M. J. Grapperhaus, J. P. Holland, and M. S. Barnes, *J. Appl. Phys.* **80**, 1337 (1996).
29. J. Hopwood, *Plasma Sources Sci. Technol.* **1**, 109 (1992).
30. M. A. Lieberman and R. A. Gottscho, in *Physics of Thin Films*, M. H. Francombe and J. L. Vossen (Eds.), Vol. 18, Academic Press, New York, 1994.

31. V. Vahedi, M. A. Lieberman, G. Dipeso, T. D. Rognlien, and D. Hewett, *J. Appl. Phys.* **78**, 1446 (1995).
32. S. M. Rossnagel and J. Hopwood, *Appl. Phys. Lett.* **63**, 3285 (1993).
33. S. M. Rossnagel and J. Hopwood, *J. Vac. Sci. Technol. B* **13**, 449 (1994).
34. M. Dickson, G. Zhong, and J. Hopwood, *J. Vac. Sci. Technol. B* **16**, 523 (1998).
35. J. E. Foster, A. E. Wendt, W. W. Wang, and J. H. Booske, *J. Vac. Sci. Technol. A* **16**, 2198 (1998).
36. M. Dickson, F. Qian, and J. Hopwood, *J. Vac. Sci. Technol. A* **15**, 340 (1997).
37. S. M. Rossnagel, *J. Vac. Sci. Technol. B* **16**, 3008 (1998).
38. S. M. Rossnagel, *J. Vac. Sci. Technol. B* **16**, 2585 (1998).
39. M. S. Barnes, J. C. Forster, and J. H. Keller, United States Patent No. 5,178,739, 1993.
40. J. E. Foster, W. W. Wang, A. E. Wendt, and J. H. Booske, *J. Vac. Sci. Technol. B* **16**, 532 (1998).
41. M. J. Grapperhaus, Z. Krivokapic, and M. J. Kushner, *J. Appl. Phys.* **83**, 35 (1998).
42. K. Suzuki, S. Okudaira, N. Sakudo, and I. Kenomato, *Jpn. J. Appl. Phys.* **16**, 1979 (1977).
43. W. M. Holber, J. S. Logan, H. J. Grabarz, J. T. C. Yeh, J. B. O. Caughman, A. Sugerman, and F. E. Turene, *J. Vac. Sci. Technol. A* **11**, 2903 (1993).
44. P. Kidd, *J. Vac. Sci. Technol. A* **9**, 466 (1991).
45. Y. Yoshida, *Appl. Phys. Lett.* **61**, 1733 (1992).
46. S. Takehiro, N. Yamanaka, H. Shindo, S. Shingubara, and Y. Horiike, *J. Appl. Phys.* **30**, 3657 (1991).
47. M. Matsuoka and K. Ono, *Appl. Phys. Lett.* **53**, 2025 (1988).
48. C. Takahashi, M. Kiuchi, T. Ono, and S. Matsuo, *J. Vac. Sci. Technol. A* **6**, 2348 (1988).
49. S. M. Gorbatkin, D. B. Poker, R. L. Rhoades, C. Doughty, L. A. Berry, and S. M. Rossnagel, *J. Vac. Sci. Technol. B* **14**, 1853 (1996).
50. J. Asmussen, T. A. Grotjohn, P.-U. Mak, and M. Perrin, *IEEE Trans. Plasma Sci.* **25**, 1196 (1997).
51. J. C. Helmer, K. E. Lai, and R. L. Anderson, Physical vapor deposition employing ion extraction from a plasma, United States Patent No. 5,482,611, 1996.
52. V. I. Miljevic, *Rev. Sci. Instrum.* **55**, 121 (1984).
53. V. I. Miljevic, *Rev. Sci. Instrum.* **67**, 1224 (1996).
54. V. I. Miljevic, *Rev. Sci. Instrum.* **69**, 1054 (1998).
55. E. M. Oks, A. V. Vizir, and G. Y. Yushkov, *Rev. Sci. Instrum.* **69**, 853 (1998).
56. A. I. Hershcovitch, V. J. Kovarik, and K. Prelec, *J. Appl. Phys.* **67**, 671 (1990).
57. A. I. Hershcovitch, *J. Appl. Phys.* **74**, 728 (1993).
58. S. Klagge and A. Lunk, *J. Appl. Phys.* **70**, 99 (1991).
59. G. Y. Yeom, J. A. Thornton, and M. J. Kushner, *J. Appl. Phys.* **65**, 3816 (1989).
60. B. Window, F. Sharples, and N. Savvides, *J. Vac. Sci. Technol. A* **4**, 196 (1986).
61. J. J. Cuomo and S. M. Rossnagel, *J. Vac. Sci. Technol. A* **4**, 393 (1986).
62. V. I. Kolobov and L. D. Tsendin, *Plasma Sources Sci. Technol.* **4**, 551 (1995).
63. Z. Dónko, *Phys. Rev. E* **57**, 7126 (1998).
64. R. R. Arslanbekov, A. A. Kudryavtsev, and R. C. Tobin, *Plasma Sources Sci. Technol.* **7**, 310 (1998).
65. A. Gunter-Schulze, *Z. Phys.* **19**, 313 (1923).
66. F. F. Chen, *Introduction to Plasma Physics and Controlled Fusion*, Plenum, New York, 1984.
67. D. B. Hayden, D. N. Ruzic, D. R. Juliano, and M. M. C. Allain, in *AVS 45th International Syposium Abstracts*, p. 179, American Vacuum Society, Baltimore, MD, 1998.

Ionization by Radio Frequency Inductively Coupled Plasma

STEVE ROSSNAGEL

T. J. Watson Research Center, IBM, Yorktown Heights, New York

I. Introduction

Ionized magnetron sputtering, or ionized physical vapor deposition (I-PVD), can be extremely useful in a number of areas relating to semiconductor interconnect technology. The mechanisms and systems used for I-PVD vary, and this book covers such topics as electron cyclotron resonance (ECR)-based ionization, hollow cathode magnetron-based ionization, and the inductively coupled radio frequency (RF) plasma-based ionization. There exist additional techniques for ionization, such as hot hollow cathode ionization, UV-photon or multiple-photon absorption ionization, electron beam ionization, which are not covered in the book but might be attractive alternatives in the future.

This chapter will examine the applications and directions in I-PVD and will be mostly based on the RF inductively coupled plasma approach. In reality, at the sample surface the origin of the ionized, depositing species, and the ionization mechanism are mostly unimportant. What matters critically is the relative flux levels of all the ionized and neutral species along with their kinetic energies.

The key applications of I-PVD in semiconductor technology rely on the intrinsic control of energy and directionality present with I-PVD. The most

Vol. 27
ISBN 0-12-533027-8
ISSN 1079-4050/00 $30.00

straightforward applications are ones in which directionality has been attained in the past by passive means of PVD, such as collimated sputtering or long-throw, low-pressure sputtering. These are slow, expensive filtering techniques which have found widespread application in semiconductor interconnect areas for bottom contact coverage, diffusion or adhesion layers, and seed layers. There are intrinsic advantages of I-PVD which make it superior to filtered sputtering from an efficiency or cost point of view. There are also opportunities with I-PVD to alter the structure of the deposited material, both topographically and microstructurally, based on the ability to control the kinetic energy of the condensing species.

II. Experimental

Most of the work described in this chapter relates to the RF inductively coupled plasma ionization approach to I-PVD.[1-3] Experimentally, a production-scale system for this is shown in Fig. 1. The magnetron is situated on the top of the tool in a sputter-down configuration. The magnetron is typically operated DC, although recent work has explored the use of moderate frequency or pulsed power for control purposes.[4] Typically, the magnetron is about 300 mm in diameter with a moving magnet array which causes the ExB etch track to cycle around the face of the cathode for high uniformity (Fig. 2). The magnetron diameter scales as approximately 1.4 to 1.5 times the sample diameter, so the transition to 300-mm Si wafers will result in magnetrons of approximately 45 cm in diameter.

The RF coils shown in Fig. 1 are generally located 3 or 4 cm vertically from the magnetron surface and have a diameter 1.2 to 1.4 times the sample

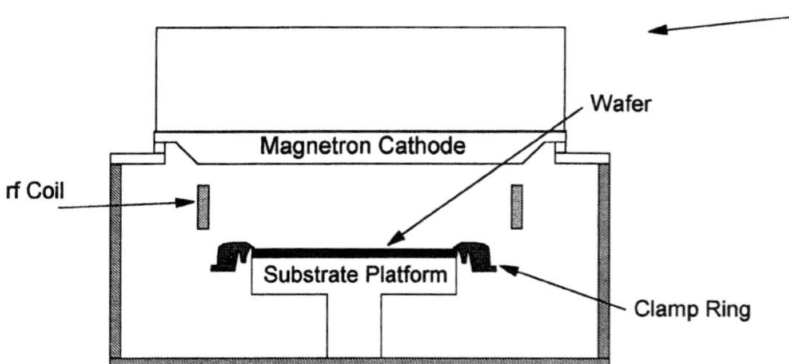

FIG. 1. Schematic diagram of an I-PVD tool based on inductively coupled RF plasma ionization of sputtered atoms.

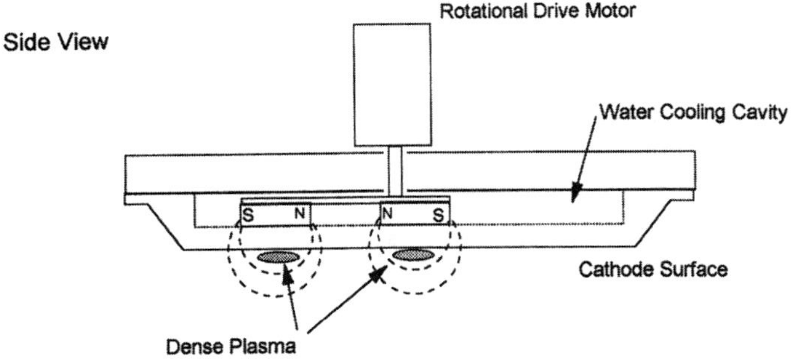

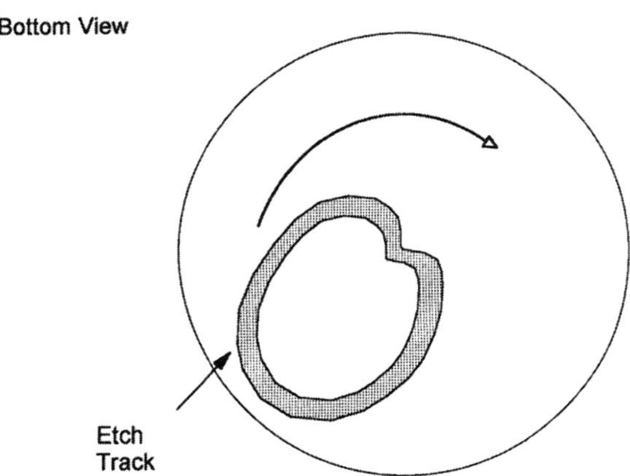

FIG. 2. Schematic of a rotating magnet magnetron cross section (top) and cathode view (bottom) (i.e., facing the sample).[5]

diameter. The coil is also located about 3–5 cm from the sample chuck and must be configured such that it does not intercept the sight line from the cathode edge to the sample edge or else it will geometrically block the outer-edge region of the sample.

The number of turns of this coil, and its construction, vary depending on the application and the experimental design. Early experimental work used RF coils of two or three turns and the coils were constructed of Cu tubing which was water cooled. Recent commercial tools (see Chapter 6) have used a single-turn coil which is not directly cooled. This approach, which uses moderate levels of RF coil sputtering to reduce film buildup on the coil, requires that the coil be constructed of the same high-purity material as that used for the magnetron cathode surface. In general, this type of coil is not easily cooled and attains a temperature of 300–500°C during operation at 1 or 2 kW of applied RF power.

The RF coil, unlike most inductively coupled RF plasma sources used for etching, is immersed in the plasma and develops a negative DC bias in much the same way as would a conventional RF-powered diode/cathode. This is due to the presence of electrons in the plasma which have much higher mobility than the ions. This negative DC bias can be controlled somewhat by tuning of the matchbox (described later) which can alter the relative level of inductive versus capacitive coupling on the coil: A higher level of inductive coupling will result in a denser plasma and a smaller DC bias on the coil. A higher plasma density will result in a larger ion current to the sample and a higher degree of ionization of the sputtered metal atoms. A higher level of capacitive coupling will result in increased RF coil sputtering and a lower plasma density and ion current to the sample.

Since the RF coil is only a few centimeters from a large metal source, namely, the magnetron, it receives a sizable metal flux and deposition rate on its surface. By altering the relative inductive and capacitive coupling levels of the coil, the result may range from net erosion of the coil with high capacitive coupling to net deposition on the coil with high inductive coupling. Since net deposition can result in concerns over eventual film peeling and particle formation, commercial tools tend to operate in a net-erosion mode. Laboratory-scale tools, often configured with water-cooled coils and perhaps less concerned with deposition and flaking, tend to operate in a net-deposition mode. This also allows the use of a variety of magnetron cathode materials without the necessity of constructing the coil from the same high-purity material.

The matching network for the inductively coupled RF plasma differs from that of a conventional RF diode, although the scale of the capacitors used is similar. The matching network is shown in Fig. 3 and consists of an initial shunt capacitor to ground and two series capacitors, one located in front of the RF coil in the plasma and one located behind it connecting it to ground. These latter two capacitors are approximately the same value (20–2000 pf, vacuum gap). The role of these capacitors is to configure the RF coil ends to be 180° out of phase, which results in the highest level of inductive coupling and the lowest DC bias on the coil.

The operating frequency used to power the RF coil is usually either 13.56 or 1.9 MHz. The former value is more readily available since it is the frequency used for the vast majority of RF plasma tools. The lower frequency is advantageous in that it is slightly more efficient in terms of plasma ionization.

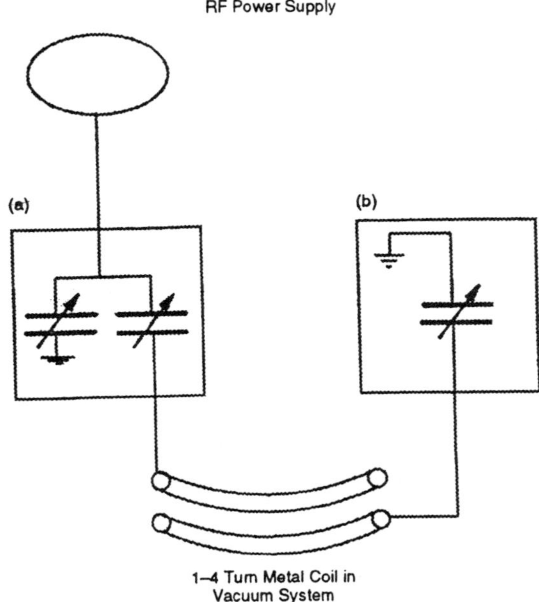

FIG. 3. Matching network used with RF inductively coupled I-PVD systems.

The sample holder in Fig. 1 is a conventional PVD sample chuck with a clamp. Although there has been a general trend away from the use of physical clamps in PVD technology, it is usually needed in I-PVD tools to maintain both good thermal contact with the sample holder and good electrical contact since there is a significant ion flux to the wafer on the order of a few amperes. The clamp ring varies from a fairly large, heavy metal ring which relies on gravity to hold the wafer down to fairly thin, spring-loaded clamps. Generally, the clamp ring is electrically connected to the substrate chuck such that any bias to the wafer is also present on the clamp ring.

The issue of edge exclusion is one which has not been well addressed with regard to I-PVD tools. Edge exclusion has significant commercial implications in terms of useable chips per wafer. In addition, for some materials systems (such as Cu/Ta) edge exclusion is mandated to reduce the chance of heavy metal contamination of the rest of the tool during wafer handling. Therefore, even though there was a distinct trend away from wafer clamping in the Al/Ti system during the past several years, clamp rings are needed with Cu/Ta and I-PVD. Some work at Novellus has examined the use of electrostatic clamping with I-PVD, and this is discussed in Chapter 5. Electrostatic clamping is ideal for wafer cooling and temperature control. For edge exclusion control with an electrostatic clamp, it is necessary to configure a shadow ring which lightly contacts the wafer edge. Conventional clamp rings which are electrically powered and hence receive a significant ion flux may heat up significantly, resulting in thermal expansion of the clamp ring and a change in the edge exclusion. It will probably be necessary to alter current clamp ring technology to include either passive cooling (between depositions) or reduced power flux, perhaps by means of ground shields or reduced clamp area exposed to the plasma.

The substrate bias can be either DC or RF, with RF bias in most reports being 13.56 MHz. Because of the presence of thick, dielectric layers on most wafer samples, RF bias is generally preferred for manufacturing applications. For a dielectric-coated wafer, using a DC bias on the clamp ring may result in significant arcing at the interface between the wafer and the ring until a continuous, conducting film is deposited. A simple fix is the deposition of a thin (50 Å) metal film on the wafer in a clampless or clamp-up configuration prior to the clamped I-PVD deposition. However, this adds an additional step to the process and is not desirable in a manufacturing situation.

The power levels for a 200-mm wafer with a clamp ring are generally 100–500 W with a substrate DC bias on the order of 40–60 V. For some samples, DC bias is adequate and similar bias voltages of −40 to −60 V result in ion currents to the sample and clamp ring assembly of 2–5 Å. Various deposition sequences may use a duty cycle of sorts for the substrate bias. Since high levels of bias can result in resputtering of the film; the deposition may be designed with a single-cycle sequence in which the bias is not used for approximately the first one-half of the deposition, and then the bias is used for the latter half, resulting in resputtering of the film deposited during the first part of the deposition. A duty cycle of 50% would indicate that the first 50% of the deposition uses no substrate bias, and the latter 50% of the deposition uses the full, stated amount of the bias.

As an alternative to the question of RF coil deposition and etching, Hopwood *et al.*[6] proposed the use of a Faraday shield located just inside the RF coils but still outside of the cathode diameter (Fig. 4). The Faraday shield requires small slits or openings to allow the RF fields to penetrate to the inner part of the chamber, and these slits must be designed to reduce deposition on insulating surfaces. The presence of the Faraday shield transfers the problem of deposition and potential peeling and flaking from the RF coils to the Faraday shield, which is typically an unpowered metal shell. The Faraday shield also functions as a sink for the diffusing metal ions in the plasma and may result in a slightly center-peaked metal plasma

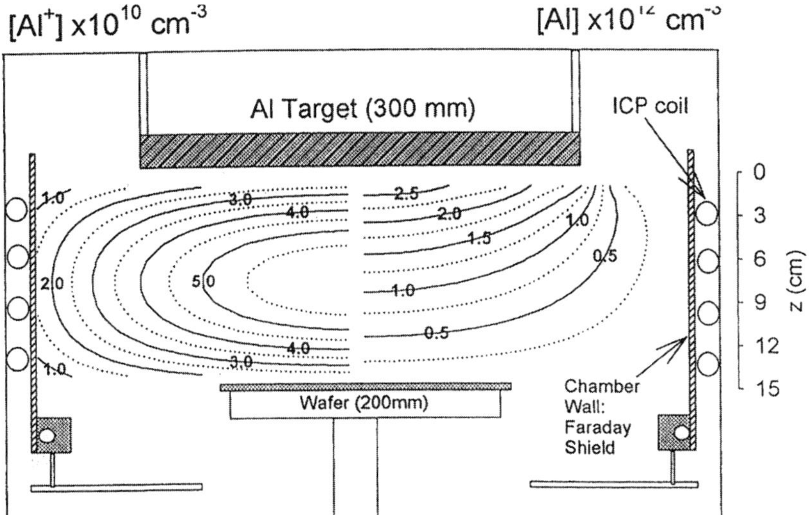

FIG. 4. I-PVD system design using Faraday shields to reduce deposition onto and sputtering of the RF coils.[6]

density. It is therefore necessary to make the shield at least 50% larger than the sample diameter.

The vacuum systems used for I-PVD are usually the same as those for conventional PVD. The chambers are usually constructed from stainless steel with a minimum number of viton O-rings. Metal conflat seals are used on most ports and feedthroughs, although O-rings are typically used for the magnetron insulator and the slit and vacuum valves. The chambers are typically cryopumped with base pressures in the 10 E−8 Torr range for most materials and 10 E−9 Torr range for Ti. Usually, the RF coil is mounted as part of a chamber spacer inserted between the chamber walls and the magnetron. This is quite similar to the case of collimated sputtering, and the spacers in both cases are 6–8 cm high.

Unlike conventional PVD or collimated PVD, I-PVD tends to operate best at pressures considerably higher than a few mTorr; typically in the 20–35-mTorr range. Since most conventional PVD chambers are unbaffled, i.e., the cryopump is exposed directly to the chamber volume when the gate valve is opened, it is necessary to baffle the pump in I-PVD to prevent warming up of the cryopump due to the high pressure and gas flow. Usually this is done by means of controlling the position of the gate valve rather than by the addition of a separate baffle. The reduction in pumping speed, when using a partially closed gate valve, must be taken into account because it increases the base pressure and the probability of impurity incorporation.

III. I-PVD Operation

Most I-PVD tools use conventional DC magnetrons as the principal metal source. The magnetron is operated in a conventional manner, typically in a power-control mode. The range of power used for a typical 300-mm diameter cathode is 500–3000 W, which is much below the design limit for these magnetrons of 20–25 kW. It is generally necessary to protect the magnetron power supply from RF pickup by the cathode. This is done by the insertion of a low-pass filter in the DC power line near the magnetron. The RF coil power supply typically is operated at 500–2500 W, and the substrate bias levels are 0–500 W.

The level of ionization in the depositing flux is the most critical feature in the use of I-PVD technology. This level, which is usually expressed as a relative ionization from 0 to 100%, is only indirectly tied to the relative level of metal ionization in the plasma. The neutral metal flux which emerges from the plasma region is simply diffusing out of the plasma region, much like a gas. The metal ions, however, are slightly pulled from the plasma edge region by a small electric field at the sheath edge (on the order of one-half the electron temperature) in a process known as Bohm presheath diffusion. This results in a significantly higher ion flux to the sheath edge than would occur simply due to gas-phase diffusion; as such, the relative ionization of the depositing flux is typically two to four times higher than the bulk ionization of metal in the plasma.

The relative ionization has been measured using gridded energy analyzers in which the collector in the analyzer is configured with a quartz crystal rate monitor.[1–3,7] This allows the analyzer to differentiate between ions of the condensable metal species and the inert gas background. Measurements have shown relative ionization levels in the depositing flux as high as 90%, with the relative ionization dependent on the level of RF power as well as the working gas pressure (Figs. 5 and 6).

It is also possible to infer the relative ionization from characteristics of the deposited film topography. The deposition onto planar surfaces will consist of contributions from both neutral and ionized metal particles. However, at the bottom of a modest aspect ratio (AR) feature, perhaps with an AR of 3 or 4, the deposition rate of neutrals is negligible and only ions will contribute to the net deposition. Therefore, for a via of AR 3 or 4, the relative ionization is simply the relative deposition rate of the bottom of the via compared to the planar areas, a ratio often called the "step coverage."

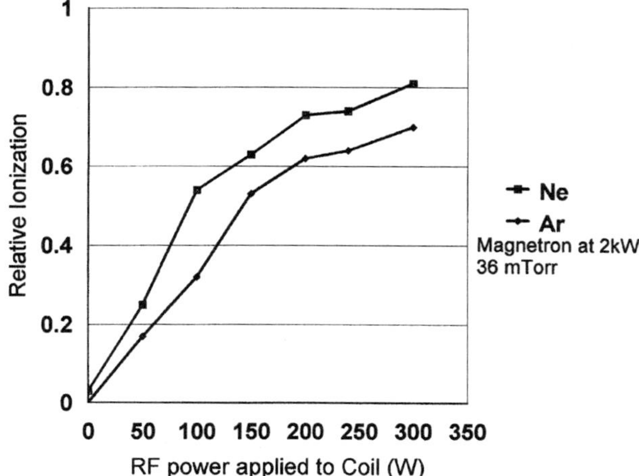

FIG. 5. Relative ionization at the sample position as a function of power applied to the RF coil.[1]

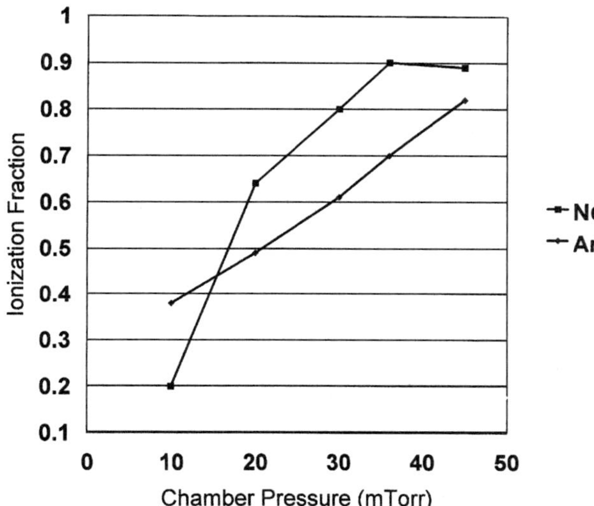

FIG. 6. Relative ionization at fixed RF power to the coil and fixed DC power to the magnetron as a function of chamber pressure.[1]

This relation is only good for thin depositions, though, since as the films build up they tend to close off the opening to the via and prevent any further deposition, whether ions are present or not.

As the AR is increased, this relationship tends to break down and the bottom step coverage is no longer equal to the relative ionization. As the AR increases above approximately 5:1, small divergences in the incident ion's trajectories, due to scattering in the sheath region and the intrinsic thermal scatter of the incident ion's velocity as it enters the sheath region, result in a declining bottom step coverage, even though the ionization level remains constant. These two effects, possibly coupled with small electrical deflections due to local charging, result in an approximately 50% reduction in bottom step coverage as the AR is increased from 5:1 to 10:1 at constant relative ionization. For example, a deposition with a relative ionization of the depositing flux of 70% results in an initial bottom step coverage of nearly 70% at an AR of 4:1, decreasing to approximately 30–35% at an AR of 10:1. The "missing" atoms are deposited on the sidewalls of the via.

The relative ionization has also been measured as a function of increasing metal levels in the plasma. The underlying assumption in the model for I-PVD is that since the metal atoms are so easily ionized compared to the inert gas, their effect on the overall plasma should be minimal, except for a slight cooling of the electron temperature and a small increase in the plasma density. However, as the metal flux increases, this assumption no longer holds. Measurement of the relative ionization of the deposited metal flux shows a significant decrease in the relative ionization as the metal flux is increased (Fig. 7). Related measurements of the ion saturation current to the sample, which is indicative of the Bohm flux (proportional to plasma density and the square root of the electron temperature), show measurable decreases in current as the metal flux is increased (Fig. 8).[8]

This situation was further complicated by measurements of the electron temperature (by Langmuir probe) and the electron density (by microwave interferometer) by Hopwood's group at Northeastern University.[9–11] These data showed a significant decrease in the electron density as the magnetron power (i.e., metal flux) was increased. They also showed very little change in the electron temperature.

This work was coupled with a plasma model, also developed by Hopwood's group, which could be used to predict the overall plasma properties in the discharge. The model suggested that the density should increase with increasing metal flux, whereas the data suggested the opposite. The paradox was resolved by allowing a modest degree of thermal heating of the background gas as a function of increased metal flux.[10,11] This conclusion was based on earlier work on gas rarefaction in conventional magnetron.[12]

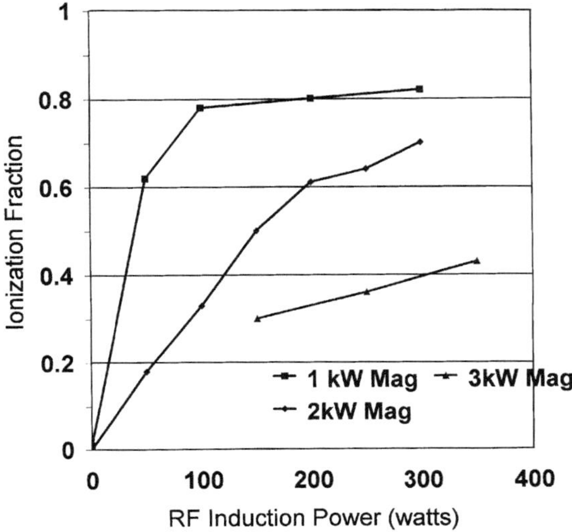

FIG. 7. Relative ionization vs RF power at three different magnetron powers.[2]

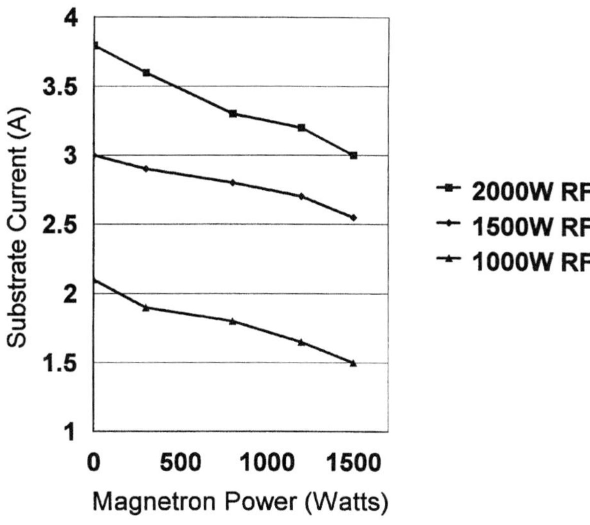

FIG. 8. Ion saturation current vs magnetron power at constant RF power.[8]

The gas rarefaction is driven by the kinetic energy of the metal atoms sputtered from the cathode, which is generally in the range of several electron volts. At low gas pressure (a few mTorr) the sputtered atoms have mean free paths which generally exceed the cathode to sample distance, so there are few in-flight collisions. However, at the 15 to 30-mTorr operating range of an I-PVD system, the majority of the sputtered atoms have in-flight collisions with Ar gas atoms, and this can result in significant energy transfer and heating of the Ar gas. It should be noted that this collisional process is indeed desired in an I-PVD tool to increase the residence time of the sputtered atoms in the plasma, resulting in a higher probability of metal ionization.

The level of gas rarefaction in a magnetron system has been experimentally measured for a conventional magnetron (Fig. 9) and also for an I-PVD system (Fig. 10).[13] For the magnetron-only case, gas rarefaction of up to 80% is observed at modest powers. For the I-PVD case, the applied RF power to the coil alone results in a density reduction of 30–55% (equivalent to the left-most points on Fig. 10). This density reduction is also due to gas heating and is a result of charge exchange and collisional heating by the dense RF plasma as well as some sputtering from the RF coils. The addition of significant metal fluxes at this point from the magnetron results in additional rarefaction of 20% or more. It is this last rarefaction effect which is responsible for the relative ionization reductions observed with I-PVD.

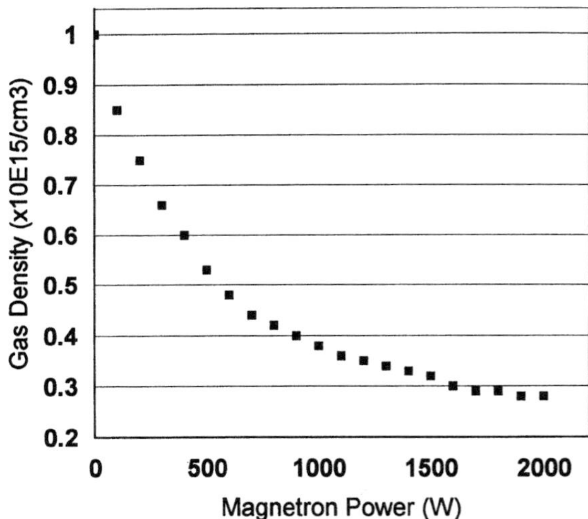

FIG. 9. Gas rarefaction for a conventional magnetron. Gas density is measured at the midplane of the system, 5.3 cm from the cathode, using a capacitance manometer.[12]

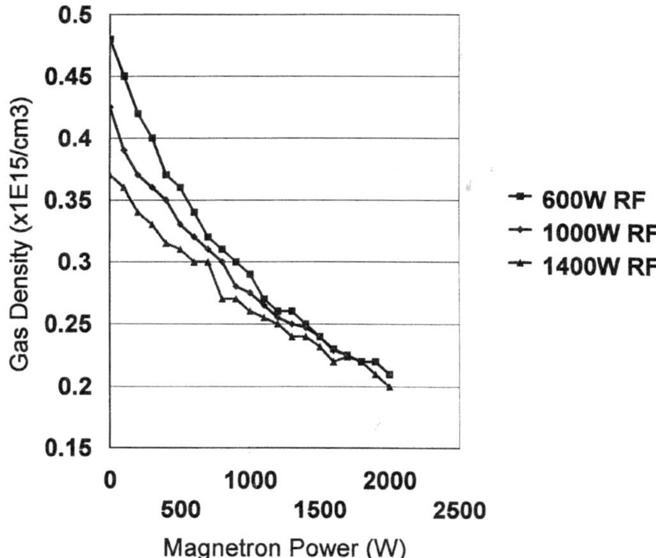

FIG. 10. I-PVD gas rarefaction: Gas density is measured at the midplane of the system, approximately 6 cm from the cathode surface, using a capacitance manometer.[13]

The following model emerges from this synergistic interplay between sputtering, gas density, transparency of the gas, and plasma density. As the metal flux to the dense plasma is increased, the sputtered metal atoms transfer heat to the background gas atoms, further rarefying the gas. As the gas is rarefied by the sputtered atom heating, the mean free path for the sputtered atoms increases and the probability of in-flight ionization decreases. In effect, increasing the metal flux results in a de facto lowering of the apparent pressure in the ionization region, making the region more transparent to the sputtered atoms. Therefore, as the metal flux is increased to the system by means of increased sputtering, fewer metal atoms have in-flight collisions in the plasma region and pass through it more rapidly (than at lower magnetron power). The result is a reduced probability for ionization of the metal atoms.

IV. I-PVD Semiconductor Applications

As will be discussed in the following sections, the desired relative ionization of the metal flux will change depending on the application. There are four primary applications for I-PVD in semiconductor fabrication: deposition primarily on the bottom of a via, a conformal coating on both the sides and the bottom of a trench or via, the filling of a trench or via, and the controlled deposition of a film with specific microstructural or chemical aspects.

A. Bottom Coverage

When vias or holes are opened in a dielectric layer on a semiconductor, the usual goal is to facilitate electrical contact with an underlying, buried circuit element, i.e., when the via is subsequently filled with a metal conductor. Generally, the top surface of the exposed, underlying circuit element is slightly oxidized, perhaps by contact with or by the deposition process for the dielectric film. Precleaning of the via bottom by either chemical means or ion bombardment may be only partly successful, depending on the metal system and the AR of the feature. One routinely used solution for W or AlCu conductors is to deposit a thin layer of Ti at the bottom of the via. The highly reactive Ti will reduce the oxide on the W or AlCu, forming a better conductive contact with the underlying metal. For modest AR vias (<2:1), collimated deposition of Ti has been adequate for deposition of a contact layer. However, for higher AR features, collimation becomes exceedingly expensive and slow for this application.

I-PVD deposition is well suited for bottom-coverage applications. The depositing ions arrive at the surface with a minimal divergence. The amount of scatter in the depositing ion flux can be predicted by estimating the maximum lateral energy due to either thermal effects or perturbations in the sheath. In either case, the divergence is related to the ratio of the lateral energy (typically <0.1 eV) to the net ion acceleration, which is the difference between the plasma potential (10–15 V) and the sample potential. This potential can vary from 0 V for a grounded sample of −15 to −20 V and an intentionally biased sample potential of many tens of volts negative. Scattering during transit across the sheath is also a factor, although the sheath thickness (0.1 mm) is generally much smaller than the mean free paths for resonant charge transfer or elastic collisions, which are closer to the millimeter scale.

The deposition at the bottom of a via of AR >3:1 is composed of >98% ions; the neutral deposition rate is almost zero. Therefore, it is possible to use a modest AR via as a diagnostic to measure the relative ionization of the depositing flux. This is most accurate for films which are much thinner than the via diameter since thicker films will tend to build out slightly at the upper corners of the via and block deposition on the bottom.

B. Conformal, Liner, or Diffusion Barrier Applications

Conformal films are generally needed to coat the inside walls and bottom of a trench or via for a chemical purpose. The applications are adhesion layers, diffusion barriers, seed layers, and possibly a layer which influences the crystal orientation of a subsequently deposited material which is used to fill up the trench or via. In this latter case, an example might be a layer which results in a preferred (111) orientation of an AlCu deposition which has greater electromigration resistance than a random orientation.

The directional nature of I-PVD appears to be incompatible with the deposition of conformal films. In practice, however, I-PVD has several components which together allow a great degree of control over the topography and conformality of the deposited layer. These components include a controllable ion-to-neutral deposition ratio and the ability to resputter the deposition film by increasing the ion energy via the sample bias.

In describing the effects of ions and neutrals, it is best to use a drawing. In Fig. 11a, the flux of depositing metal ions results in essentially bottom-only coverage. In Fig. 11b, the neutral flux is shown, which is mostly isotropic and results in deposition primarily near the top corners of the deposition. The sum of the depositions is shown in Fig. 11c, and this has a profile quite similar to a collimated deposition.

This type of deposition profile is adequate for some applications for diffusion barriers and liners, seed, and adhesion layers. However, much like collimation, the weak point is at the bottom corners of the via or trench where there is a seam or crack between the dense bottom deposit and the much less dense sidewall deposit. This is adequate for some applications, such as Ta and TaN, in which the depositing atoms have a less than unity sticking probability and hence tend to smear out the nonconformality of the deposit. It is much less applicable for Ti, TiN, and Cu.

A straightforward solution is to increase the kinetic energy of the depositing ions (and the inert gas ions which impact the sample) by making the sample bias more negative. The plasma potential, which is the electrical starting place for the depositing ions, is unaffected by the sample potential, and the net ion energy is the difference between the plasma potential and the sample potential. Since RF bias is involved in most commercial applications of I-PVD, the sample bias is controlled by adjusting the RF power supplied to the sample pedestal. Typically, for most materials used with I-PVD, a sample bias of -40 V or more leads to significant resputtering of the deposited films. An approximate RF power level equivalent to -40 to -50 V for a 200-mm pedestal is approximately 300 W. It should be

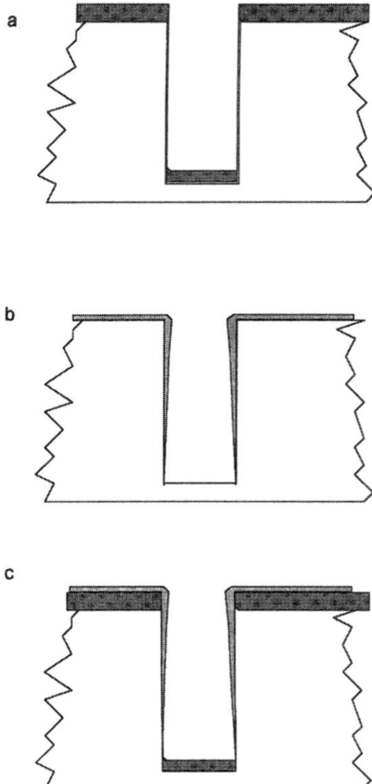

FIG. 11. (a) Deposition due to ions only, (b) deposition due to neutrals only, and (c) net deposition with no resputtering.

noted, though, that sample bias of this magnitude or greater can result in significant compressive stress in the deposited films.

The result of the increased ion energy is a resputtering of the depositing film from the horizontal surfaces, i.e., surfaces parallel to the plane of the sample. Three primary effects occur: (i) Atoms are sputtered from the field regions of the wafer, resulting in a lower net deposition rate on the field; (ii) edges, such as those at the top corners of trenches or vias, may be beveled due to the higher sputter yield of many materials at a 40–60° angle of incidence; and (iii) atoms may be sputtered from the bottoms of bias and trenches. These effects are shown in Fig. 12.

The key effect for near-conformal film deposition within features is the third item listed previously, the resputtering from the bottom of the feature. Virtually all these sputtered atoms are recaptured by the sidewalls of the

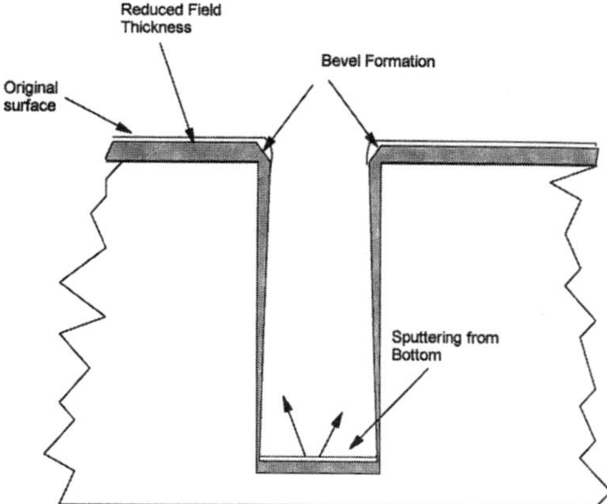

FIG. 12. I-PVD deposition showing sputtering of field region by energetic ions, beveling of corners, and sputtering from the bottoms of features.

feature, with the highest deposition rate at the bottom corner. The result (Fig. 13) is that the atoms deposited on the bottom of the feature can be moved to the lower sidewalls depending on the amount of resputtering. Small amounts of resputtering cause a slight corner thickening.[14] The other extreme is the complete removal of the bottom deposit, resulting in significant thickening of the bottom corners.

It should be noted that the term "step coverage" is distorted in this type of scenario because the field thickness, which is the denominator in the step coverage calculation, is decreased by the resputtering. Therefore, it is possible to have well over 100% step coverage for the sidewall films, although by itself this number has relatively little meaning.

Bevel formation at the top corners of the feature is a significant problem for filling applications, but it has relatively little effect for thin liner films because the bevel is on the order of the film thickness and for very thin films it is not critical. It is important, though, that the bevel not reach the corner of the dielectric, exposing it and increasing the net feature size at the field level. This limits resputtering effects to most likely less than unity sputter yields, and it may limit applications in which it is desired to remove entirely the bottom coverage.

Resputtering effects are often controlled by a duty cycle. At low ion energy (low or no bias or applied RF power), the deposit is mostly directional resulting in good bottom coverage as well as coverage of the fragile upper corners. After most of the desired film thickness is deposited, the sample bias is turned on to a level which can cause significant resputtering (several hundred watts). While this results in a effective sputter yield which is >1, the upper corners are well covered and the most notable effect occurs with the deposit at the bottom of the trench or via, where the deposition from the first part of the cycle is moved to the sidewalls. This can be described as a single-cycle "duty cycle" and is usually described in terms of the percentage of time that the substrate power is energized, on the order of 40–75%. Obviously, more complicated duty cycles are possible with more than a single cycle.

This combination process can be tailored to the physical dimensions of the desired feature. An example is shown for features of AR 5:1 with varying degrees of resputtering and ionization (Fig. 14). Computer modeling suggests that sidewall coverage at the 5% level is possible up to about an AR of 5:1.[15] At the 10:1 level, it is debatable whether there is a continuous film down the sidewall. Also, effects such as surface diffusion and agglomeration which are common for Cu may alter the net deposit or topography.

C. Filling of Trenches and Vias

Perhaps the most important component of metallization for interconnects is the primary conductor, which has historically been AlCu but Cu and perhaps dilute Cu alloys are increasingly being used. At the junction level, the first via is usually made from CVD W, and this is likely not to change in the evolution from AlCu to Cu. This via typically requires a contact Ti layer and then a conformal TiN diffusion barrier. In the early to mid-1990s, these layers were usually deposited by means of collimated sputtering. In the late 1990s, many of the collimated PVD systems are being upgraded to I-PVD due to decreased cost of ownership and increased reliability.

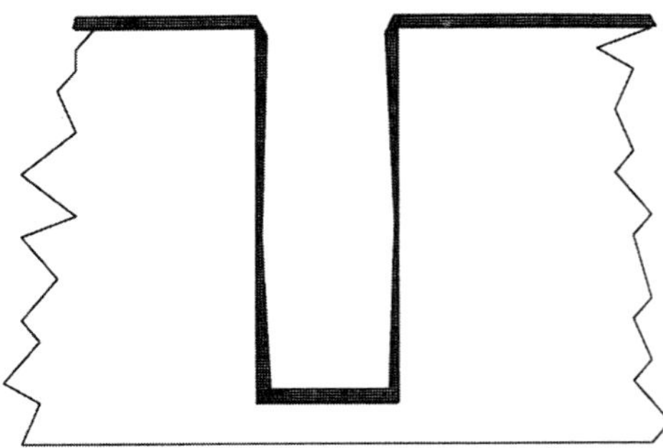

FIG. 13. The effects of resputtering and redeposition on the films within a via or trench.[14]

For patterning techniques based on RIE (Reactive Ion Etching) of planar AlCu layers, conventional PVD was adequate. With the evolution of damascene processing, a number of other PVD, chemical vapor deposition (CVD) and electroplating variants have been explored. One or more of these techniques is needed to overcome the intrinsic flaw of conventional PVD:

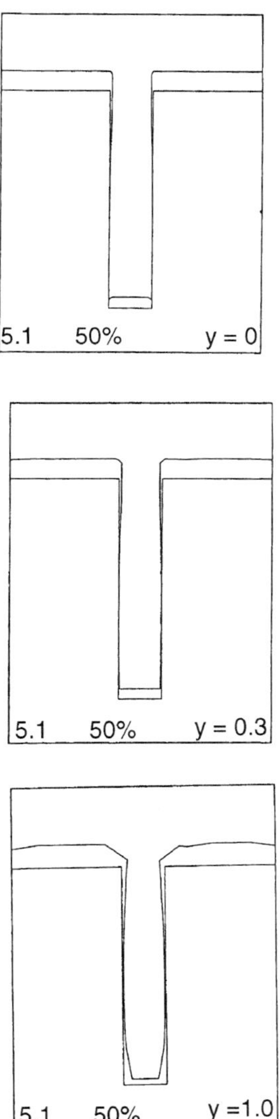

FIG. 14. Films deposited in 5:1 AR features.

rapid void formation when depositing films into features with an AR greater than about 0.5:1. This chapter will not explore either CVD or electroplating technology and will focus primarily on I-PVD technology with various connections to other PVD techniques, such as collimation and reflow.

The requirements for filling for the $<0.2\,\mu$m semiconductor generations are fairly severe. The ARs range from 3 to 4:1 for microprocessor applications to nearly 8:1 for DRAM at the 1-gigabit level and beyond. Sidewall angles are approximately 90° and the thermal budget of the wafer is declining with each generation. Currently, the maximum wafer temperature is constrained to 400°C, which is required for damage annealing and also for some wafer or wire-bonding steps. The required development of low k dielectrics may reduce this thermal budget very significantly, perhaps by several hundred degrees in a few generations. Clearly, processing steps which require highly elevated temperatures should not be developed since they are completely counter to this overall direction.

I-PVD has several advantages and disadvantages when applied to filling applications. The primary advantage is that the directionality of the deposit is derived inherently from the process: Ions are directed at 90° toward the sample. This is independent of sample temperature and appears to work for all the commonly used semiconductor materials. In addition, PVD-based technologies such as I-PVD are consistent with the existing, well-engineered PVD toolset which is present throughout the semiconductor industry. PVD is a well-known, well-documented technology with broad acceptance and presence in the industry.

The principal disadvantage of I-PVD as it is currently practiced is that the ionization level of the depositing flux is not 100% but more typically 50–70% depending on the operating conditions. Higher ionization levels approaching 90% have been reported in experimental systems, but these high levels do not scale up well in terms of sputtering and deposition rate due to the rarefaction process described previously.

The result of $<100\%$ ionization is shown in Fig. 15. The deposition looks quite similar to a collimated PVD deposition, in which there is a relatively high-density bottom deposit and a columnar, low-density deposition on the sidewalls, with the two regions separated by a seam. When polished [using chemical–mechanical processing (CMP)] back to the dielectric surface, the seams are quite evident on the metal surface. Interconnect lines such as these are likely to have very poor electromigration resistance because of the presence of the seam or void which is aligned with the direction of current flow. The level of bottom filling for modest AR features is shown in Fig. 16.[8]

It is clear that some degree of atomic mobility is required in this case to

(i) remove the seam between the bottom and sidewall deposits and densify the sidewall films and to (ii) remove or thin the thick deposits on the field which require subsequent CMP to planarize. If the depositing flux were 100% ionized, the seam between the two deposited films might be minimized or at least transferred very close to the sidewall. However, this would not eliminate the thick deposit on the field, which would have to be as thick as the deepest feature on the wafer is deep.

Surface diffusion of Cu on Cu is well documented, in particular for PVD Cu depositions by D. Gardner of Intel.[16,17] Also known generically as "reflow," the techniques for PVD diffusion-based filling involve temperatures in the 350–450°C range, very clean vacuum systems (<10 E–8 T), usually the presence of a seed layer which is well bonded and conformal, and adequate time to allow the diffusing atoms to migrate into the trenches and vias. The situation with Cu is actually similar to that of reflow of Al metal. There is debate over the relative value of PVD seed layers as opposed to CVD layers, but there are few definitive experiments.

An I-PVD-based reflow scheme has two intrinsic advantages over CVD-based techniques and even conventional PVD schemes. First, the I-PVD chamber can be used for the cold seed layer, in much the same way that it is used for seed layers for electroplating. A second deposition step can then use either PVD or I-PVD at elevated temperature to facilitate the filling process. Second, there is a mild degree of bombardment-enhancement to the surface diffusion process. This depends greatly on the materials system used, but this may result in a lowering of the required temperature needed for reflow, reducing the overall thermal budget of the wafer. Little has been published in this area recently, and it remains an open field of interest.

One of the unique aspects of I-PVD, i.e., the ability to self-sputter the depositing film, is not very useful for the filling of modest to high AR features. The goal behind increased self-sputtering of the deposit film is to reduce the overhang formation at the top corners of the deposit, resulting in a more open feature. This works moderately well for low AR features (Fig. 17), and even though the net AR of the feature appears to increase with thickness the feature is completely filled.

At higher ARs, though, the effect of bevel formation at the top corners of the deposit can be fatal. The problem is not the bevel formation itself but the redeposition of material which is sputtered from the bevel surface and lands on the opposite sidewall. This was shown first by experiment and then by modeling by several authors.[18,19] The result is shown in Fig. 18 for an AR of only 2:1. As the level of resputtering is increased (which is equivalent to increasing the net sputter yield of the film), large bevels form, which at high levels of resputtering close off the via leaving behind a large void.

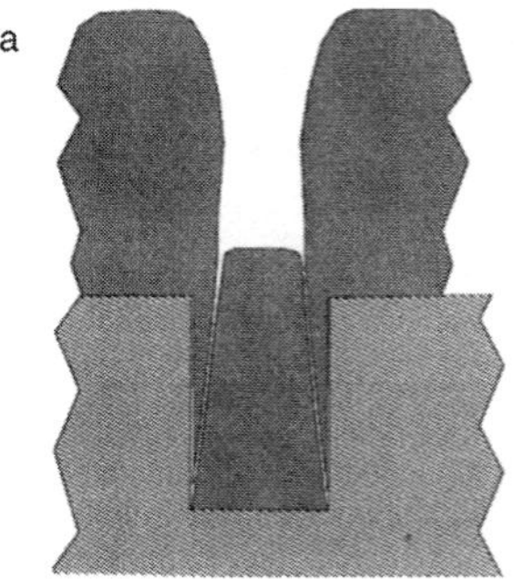

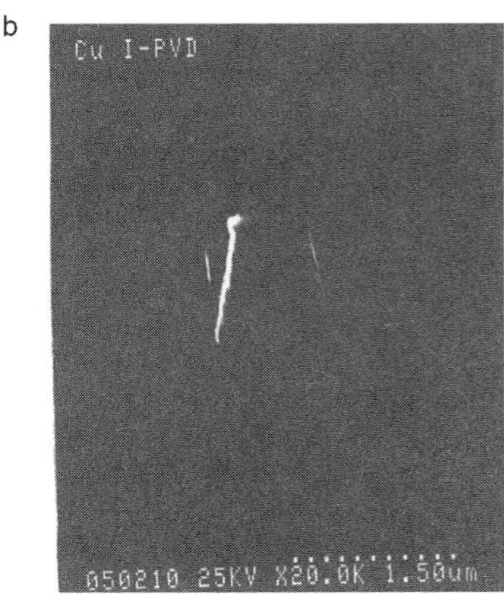

FIG. 15. (a) Sketch of as-deposited I-PVD deposition with 50% relative ionization of the depositing flux. (b) SEM of feature after polishing.

D. MICROSTRUCTURAL AND CHEMICAL ASPECTS OF I-PVD

One of the unique aspects of depositing films from mostly ions is that the kinetic energy of the ions can be easily controlled. This has been done in the past in the form of partially ionized evaporation or ion plating, but it has not been readily available for PVD-based films. Generally, in past studies of the effect of extra kinetic energy deposited during deposition, the trend was to use a broad-beam ion source with an Ar, oxygen, or nitrogen beam which was incident on the growing film concurrently with the deposition process.[20] This work resulted in the observation of significant changes in film crystallinity or stoichiometry due to the ion bombardment, and this was usually characterized in the form of "eV per deposited atom," where the energy was due to inert gas ion bombardment and the depositing atom was both neutral

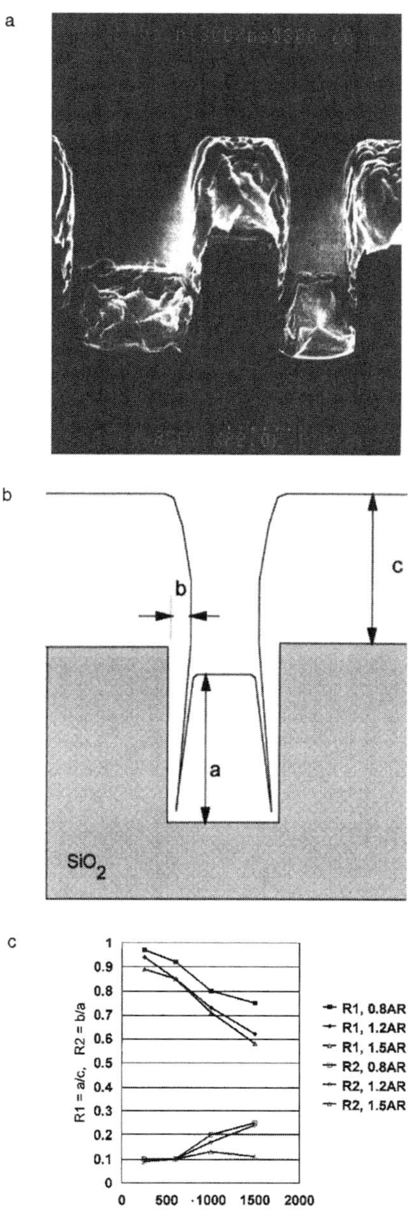

FIG. 16. (a) SEM of as-deposited I-PVD Cu with 50% ionization, (b) sketch of deposit showing measurements, and (c) plot of bottom step coverage (R^2 = a/c) and relative sidewall coverage (R^2 = b/a) as a function of magnetron power at various RF powers.[8]

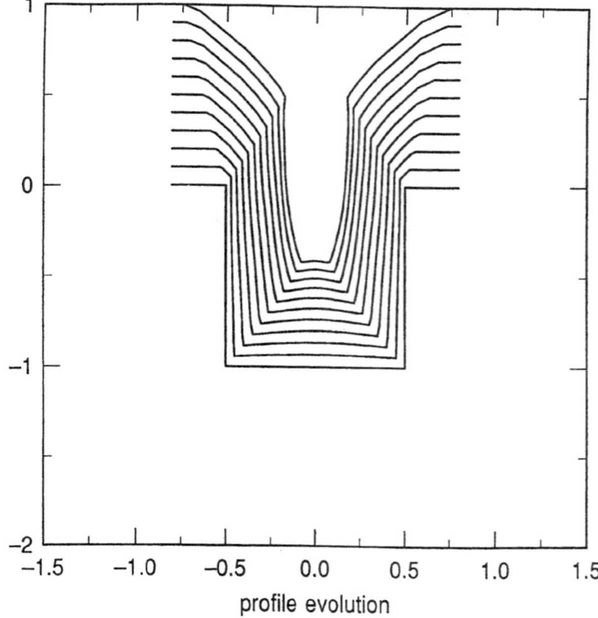

FIG. 17. The effect of resputtering of a 1:1 AR feature with 50% ionization and a sputter yield of 0.5.

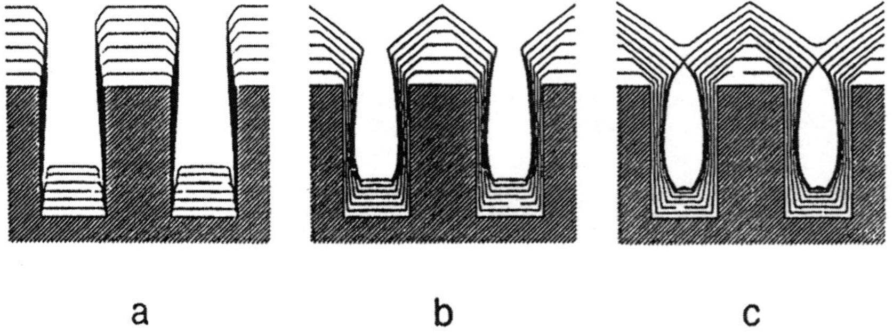

FIG. 18. The effect of resputtering on filling. (a) Resputter yield = 0, (b) resputter yield = 0.3, (c) resputter yield = 1.0. For each case, the relative ionization of the depositing film is 50%.[18]

and rather low energy. This sort of energy-averaging process was mildly successful in characterizing the effects of bombardment-enhanced deposition, with most effects dependent on energies of perhaps 10–40 eV/depositing atom.

I-PVD techniques can result in high relative ionization levels for the depositing flux (as high as 90% ions). Therefore, the vast majority of the condensing atoms can be deposited under controlled-energy circumstances. However, there is also a significant Ar^+ flux to the sample, which is typically two or three times the metal ion flux. This is because the plasma in the chamber is still mostly dominated by the inert gas species: The metal is ideally only a perturbation to the high-density inert gas plasma. Therefore, the real energy deposited per condensing atoms can be two to four times higher than simply the bias potential plus the plasma potential. Nevertheless, it is still relevant to examine changes in film chemistry or structure which relate to the kinetic energy of the ions (both gaseous and metal) which impact the film surface.

1. Film Stoichiometry

Reactive PVD is generally practiced by the sputtering of a metallic target in a gas mixture of an inert gas species (Ar) and a reactive gas, such as oxygen or nitrogen. This process is complicated by chemical changes that may occur on the target surface. For example, when adding oxygen in the case of Ar sputtering of Al, when there is sufficient oxygen in the system to form stoichiometric alumina films the Al target is also oxidized, which results in a severe drop-off (20 times) of the sputter yield and even greater oxidation of the Al target. The nitride cases are slightly more forgiving since the sputter yield difference from the pure metal case to the nitride phase is small. However, the problem remains that it is difficult to control these processes. In addition, for the nitride case, the reactions are often not spontaneous and require either significant sample heating or substrate bias in some form to help drive the reaction.

Reactive I-PVD has been applied mostly to the case of TiN. In this case, a Ti cathode is used with Ar gas and a partial pressure of nitrogen. Two results occur which are different from those of conventional reactive-PVD case. First, the amount of nitrogen needed to form stoichiometric TiN films is significantly reduced compared to that needed for the conventional PVD case (as much as two or three times). Second, the formation of TiN is clearly dependent on the kinetic energy of the ions, i.e., the sample bias potential. This is shown in Fig. 19, which plots resistively of the TiN films as a function of sample vias.

These results suggest that the kinetic energy from the depositing (Ti and N) ions and the noncondensing (Ar) ions can be helpful in driving the nitride reaction on the sample surface. It should be noted that these results were obtained at room temperature, so the primary energy input was from the ions. Another aspect of this reactive deposition case is that the mag-

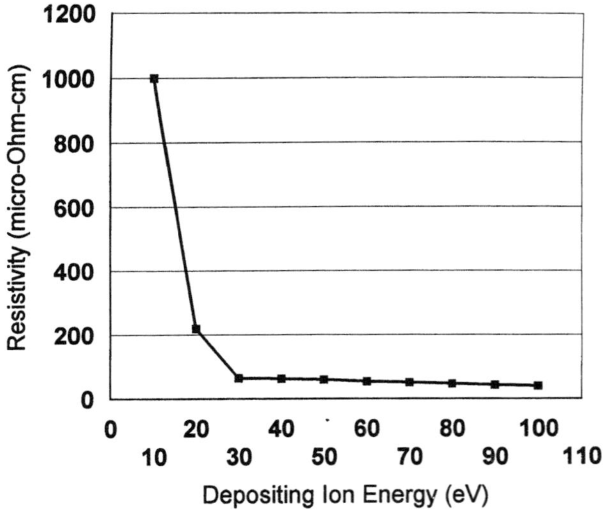

FIG. 19. TiN resistivity for 1000 Å films as a function of sample bias potential. The plasma potential in each case is approximately +10 V.[21]

netron cathode does not experience a significant change in operating voltage or deposition rate due to the presence of the reactive species. This is partly due to the much-reduced level of reactive gas in the chamber and may also be due to the triode nature of the magnetron cathode in the I-PVD case: The discharge is dominated more by the inductively coupled RF plasma than by the magnetron-generated plasma. This area remains open for additional work and understanding.

2. *Film Crystallinity*

It is not surprising that the degree of crystallinity of a film and perhaps its orientation can be altered by the addition of energy to the depositing film. In many ways, added kinetic energy to the depositing ions will have similar results to those obtained with an increase in the substrate temperature: Both will result in some degree of additional adatom mobility, grain growth, and perhaps the facilitation of a preferred crystalline orientation.

There have been few studies of the effect of ion energy on film crystallinity during I-PVD. The earliest work was with AlCu films,[22,23] and this work compared I-PVD deposition with careful ionized-beam depositions in similar circumstances. The general result was that the crystallinity could be

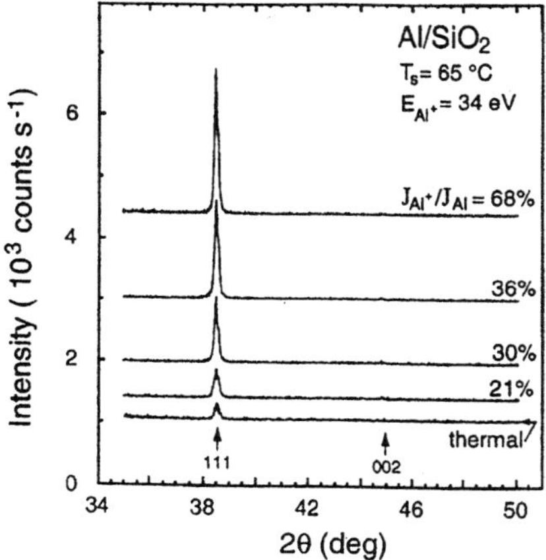

FIG. 20. An XRD 2-θ trace of I-PVD Al films deposited on silicon dioxide with an ion energy of 34 eV for various ionization levels.[22]

controllably altered by increasing the kinetic energy of the depositing metal atoms. For AlCu, this leads to a preferred (111) orientation (on silicon dioxide) with a degree of crystallinity which is dependent on either the relative ionization or the ion energy (Figs. 20 and 21). In addition, the orientation of the films, as measured by the full width at half-maximum (FWHM) of the X-ray peaks, was sharpened considerably as the ion energy was increased (Fig. 22).

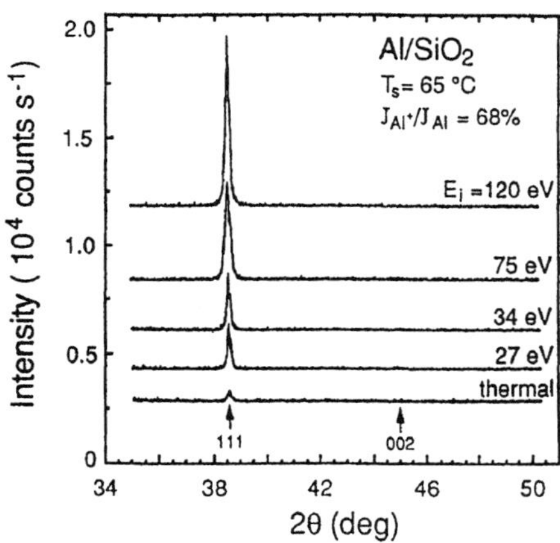

FIG. 21. An XRD 2θ trace of I-PVD Al films deposited on silicon dioxide with a constant ionization fraction of 68% as a function of ion energy.[22]

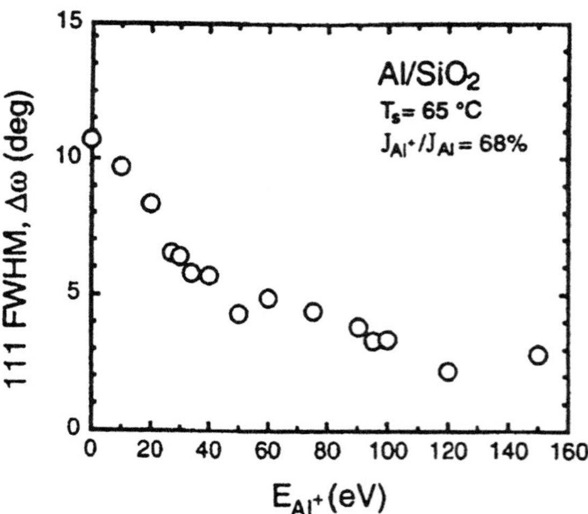

FIG. 22. The FWHM for the (111) Al peak for I-PVD Al deposited on silicon dioxide for a constant ionization fraction of 68% as a function of ion energy.[22]

In a completely different system, exploring the degree of crystallinity of reactively sputtered I-PVD of alumina, workers at Northwestern University observed changes in the temperature at which certain phases were formed. For the "kappa" phase of alumina, which is generally considered the hardest phase, the degree of crystallinity was observed at significantly lower substrate temperature (500°C) than would be observed by simply heating the substrate alone (900°C).[24]

V. Conclusions and Future Directions

I-PVD is a technology which has significant promise for the semiconductor industry and for related thin film applications. The unique advantage of I-PVD is that, on a significant scale, films can be deposited primarily from ions rather than neutrals. The main advantage of ion deposition is that the direction and energy of the condensing particles can be controlled electrically, which has not been possible with past techniques which were based on neutral atom deposition. The control of direction allows the application of projecting the depositing ions into high AR features, which fundamentally alters the nature of PVD depositions. The control of energy allows great control over both the physical and the chemical nature of the deposited films as well as the ability to selectively resputter parts of the film and rearrange the deposition topography in a controlled manner.

The specific type of I-PVD, whether it is based on ECR-based ionization, indictively coupled RF ionization, hollow cathode ionization, or some other technique, is mostly irrelevant. Even the source of the condensing particles is mostly irrelevant, and the earliest work used evaporation rather than sputtering as the metal source. The real value of the generic I-PVD technique is the intrinsic control over the deposition which is derived from the sensitivity of ions to electric fields, which is completely foreign to conventional neutral deposition technologies.

References

1. S. M. Rossnagel and J. Hopwood, *Appl. Phys. Lett.* **63**, 2903 (1993).
2. S. M. Rossnagel and J. Hopwood, *J. Vac. Sci. Technol. B* **12**, 449 (1994).
3. S. M. Rossnagel, *Thin Solid Films* **263**, 1 (1995).
4. Jung Hoon Joo, paper presented at American Vacuum Society Annual Symposium, San Jose CA, October 1997.
5. R. A. Powell and S. M. Rossnagel, in *PVD for Microelectronics*, Academic Press, Boston, 1998.
6. M. Dickson, G. Zhong, and J. Hopwood, *J. Vac. Sci Technol. B* **16**, 523 (1998).
7. D. B. Hayden, D. R. Juliano, K. M. Green, D. Ruzic, C. A. Weiss, K. A. Ashtiani, and T. L. Licata, *J. Vac. Sci. Technol. A* **16**, 624 (1998).
8. C. A. Nichols, S. M. Rossnagel, and S. Hamaguchi, *J. Vac. Sci. Technol. B* **14**, 3270 (1996).
9. J. Jopwood and F. Qian, *J. Appl. Phys.* **78**, 576 (1995).
10. M. Dickson, F. Qian, and J. Hopwood, *J. Vac. Sci. Technol. A* **15**, 340 (1997).
11. M. Dickson and J. Hopwood, *J. Vac. Sci. Technol. A* **15**, 2307 (1997).
12. S. M. Rossnagel, *J. Vac. Sci. Technol. A* **6**, 19 (1988).
13. S. M. Rossnagel, *J. Vac. Sci. Technol. B* **16**, 3008 (1998).
14. S. Hamaguchi and S. M. Rossnagel, *J. Vac. Sci. Technol. B* **14**, 2603 (1996).
15. S. M. Rossnagel, paper presented at American Vacuum Society Annual Symposium, San Jose, CA, October 1997.
16. D. S. Gardner and D. B. Fraser, Proceedings of the VLSI Multilevel Metallization Conference (VMIC), Santa Clara CA, June 27, 1995, p. 287.
17. L. J. Friedrich, D. S. Gardner, S. K. Dew, M. J. Brett, and T. Smy, *J. Vac. Sci. Technol. B* **15**, 1780 (1997).
18. S. Hamaguchi and S. M. Rossnagel, *J. Vac. Sci. Technol. B* **13**, 125 (1995).
19. T. S. Cale and V. Mahadev, in *Modeling of Film Deposition for Microelectronic Applications* (S. M. Rossnagel, Ed.), p. 176, Academic Press, Boston, 1996.
20. E. Kay and S. M. Rossnagel, in *Handbook of Ion Beam Processing Technology* (J. Cuomo, S. Rossnagel, and H. Kaufman, Eds.), p. 170, Noyes, Park Ridge, NJ, 1989.
21. S. M. Rossnagel, in *Beam–Solid Interactions for Materials Synthesis and Characterization* (D. E. Luzzi, T. F. Heinz, M. Iwaki, and D. C. Jacobson, Eds.), Vol. 354, Materials Research Society, Pittsburgh, 1995.

22. Y.-W. Kim, J. Moser, I. Petrov, J. E. Greene, and S. M. Rossnagel, *J. Vac. Sci. Technol. A* **12**, 3169 (1994).
23. Y.-W. Kim, I. Petrov, J. E. Greene, and S. M. Rossnagel, *J. Vac. Sci. Technol. A* **14**, 356 (1995).
24. J. M. Schneider, W. D. Sproul, A. A. Voevodin, and A. Matthews, *J. Vac. Sci. Technol. A* **15**, 1084 (1997).

Ionization by Microwave Electron Cyclotron Resonance Plasma

William M. Holber

Applied Science and Technology, Inc. Woburn, Massachusetts

I. Introduction

Electron cyclotron resonance (ECR) plasma discharges were first systematically studied as part of several plasma fusion programs, beginning in the 1960s.[1] Use of ECR plasmas for materials processing began in the 1970s.[2–4] By the mid-1980s, ECR plasma reactors were being used in a wide range of research and early production activities, primarily in the etch area. By the late 1980s activity extended to a wide variety of deposition processes.

This section will begin with a general review of the principles of ECR plasma generation. It will then provide a broad background as to the uses of ECR plasmas in materials processing, followed by an overview of the literature on ECR-based ionized physical vapor deposition (I-PVD) activity. Several applications in copper and aluminum deposition for microelectronics will be discussed in detail.

ECR plasmas are generated through the interaction of microwave radiation and a magnetic field set to a value such that the absorption of the microwave energy by the electrons in the plasma is resonant. While the basic theory is quite simple and easy to understand, there are complexities beyond the basic theory which will not be discussed here. Numerous books and articles[5–7] are available for those interested.

An electron in motion in a uniform, static magnetic field will undergo circular motion tranverse to the magnetic field direction with frequency (the

Vol. 27
ISBN 0-12-533027-8

ISSN 1079-4050/00 $30.00

cyclotron frequency) given in S.I. units by $\omega_c = eB/m$, where e is the electron charge, B is the magnetic field strength, and m is the mass of the electron. When an electromagnetic field is applied, energy can be transferred from the field to the electrons. A resonance condition exists when the electron undergoes one circular orbit in one period of the applied electromagnetic field.

For arbitrary direction of propagation of the microwave radiation relative to the magnetic field, the electromagnetic radiation may not be able to penetrate the bulk of the plasma once the plasma has reached a critical density, given in S.I. units by $N_c = (m\varepsilon_0\omega^2)/e^2$, where ω is the frequency of the electromagnetic wave. In such cases, the bulk plasma density can be increased by raising the frequency of the electromagnetic field—and the power as well. For some plasma fusion experiments, frequencies in excess of 100 GHz have been studied.

Unfortunately, high-frequency microwave usage is not practical for most materials processing applications due to the cost of the power at that frequency and the size and expense of the correspondingly stronger magnets needed to create the resonant magnetic field. Most materials processing applications use microwaves at 2.45 GHz for the following reasons: (i) 2.45 GHz is reserved by international agreements for industrial applications, (ii) hardware and power supplies are readily available and relatively inexpensive since it is an industrial standard frequency and also the frequency used for most microwave heating applications, and (iii) the resonant magnetic field strength of 875 Gauss can be achieved with either electromagnets or permanent magnets having acceptable cost and size parameters.

The critical plasma density for 2.45-GHz microwave excitation is $7 \times 10^{10}\,\text{cm}^{-3}$. It is possible under specific conditions of microwave launch into the plasma to obtain an overdense plasma, which can have a bulk plasma density much greater than this value. This is an important design basis for ECR plasma reactors. The criteria which must be fulfilled in order to achieve overdense plasma operation follow directly from the simple dispersion theory of magnetized plasmas and are:

1. The microwave radiation is launched into the plasma in an orientation such that its direction of propagation is parallel or near parallel to an axial magnetic field. The electromagnetic wave is right-hand circularly polarized relative to the magnetic field vector.
2. The magnetic field has a gradient from the launch point of the microwave into the vacuum chamber such that its (scalar) value from the launch point to the point at which it is resonant is always greater than the resonant value.

Numerous works[8–14] discuss both the theory and the basic physics measurements for ECR materials processing plasmas. To summarize, using 2.45-GHz microwave power plasma, densities $>10^{12}\,\mathrm{cm}^{-3}$ can be obtained when an optimum geometry is employed. Plasmas can be generated uniformly over diameters of >200 mm through careful design of both the source and the magnetic field profile. The typical background pressure used is in the 1 to 5 mTorr range, although ECR plasmas can be generated and sustained at pressures less than 0.5 mTorr and more than 10 mTorr. For optimally designed systems, plasma density scales linearly with increasing microwave power.

Numerous varieties of ECR plasma reactors have been constructed. In the divergent magnetic field ECR reactor the magnetic field lines diverge from the resonant point to the substrate. Often, electromagnets are used both in the source region (where the microwave is launched and absorbed) and downstream in the vicinity of the substrate being processed. The primary role of the source electromagnets is to set the conditions for the electromagnetic wave launch and absorption. The primary role of the downstream electromagnet is to control the divergence of the plasma so as to adjust the plasma uniformity at the substrate. Typical operating parameters reported are microwave powers of 1–5 kW, background pressures of 1–5 mTorr, and plasma densities at the substrate of $3–8 \times 10^{11}\,\mathrm{cm}^{-3}$.

Divergent magnetic field ECR reactors which utilize permanent magnets instead of electromagnets have also been constructed.[15,16] This type of design offers compactness and operational simplicity relative to the electromagnet version, though at some cost in flexibility. The downstream permanent magnets are set in a multipole array which provides some confinement for the electrons at the chamber wall while allowing for nearly zero magnetic field strength over the process substrate. This promotes plasma diffusion, improving plasma uniformity over the substrate.

ECR reactors have been used for a wide variety of materials processing, including etch, deposition, surface cleaning, and surface modification. Most of these studies have employed conventional gaseous reactants or precursors: for example, Cl_2, HBr, and C_2F_6/O_2 in the case of etch and SiH_4 and SiH_4/O_2 in the case of deposition. ECR deposition studies using gaseous precursors include a-Si,[17–20] poly-Si,[21,22] SiO_2 and Si_3N_4,[23–31] BN and B_xO,[32–34] TiN,[35] DLC,[36] and PZT.[37]

In the area of ECR I-PVD, in which the deposition precursors are nonvolatile and a large fraction of the deposited material is ionized by passage through the ECR-generated plasma, much of the work to date has focused on specific applications in silicon semiconductor production, driven by the need to deposit metal into high aspect ratio interconnect structures,[38–46] including Al[39,45,47–50] and Cu.[38,40–44,46,51,52] There has been a large body of other work as well, including poly-Si,[53] Al_2O_3,[51–56] ZnO,[49,57] Mo,[47–49] Fe,[47–49,58] $BaTiO_3$,[49] TiN,[50] $BiTi_3O_{12}$,[59] TiNi,[60,61] BN,[62] and $YBa_2Cu_3O_{7-x}$.[63]

II. Techniques in Generating ECR I-PVD Plasmas

Generation of an ECR I-PVD plasma involves physics issues not evident in purely gas-source reactors. In gas-source reactors, the reactant fed into the plasma is composed of volatile species, such as SiH_4 and Cl_2. The concentration of species equilibrates throughout the plasma reactor. Most of the molecules and atoms which hit the chamber wall generally are also volatile so that they recycle into the plasma where they can be reionized or dissociated.

In ECR I-PVD reactors the active species are generally not volatile. Therefore, there are new issues of how to introduce the reactant into the plasma and what happens as those species are lost via collisions with walls and other chamber surfaces on which they stick. Both evaporation and sputtering techniques have been used for reactant introduction.

Evaporation can be used to introduce elemental species into the plasma. Work reported includes evaporation of aluminum and copper,[38] coevaporation of Y, Ba, and Cu to deposit $YBa_2Cu_3O_{7-x}$,[63] and evaporation into an oxygen plasma for deposition of Al_2O_3.[55] The advantages of evaporation as a technique are (i) the evaporant feed can be extremely pure, (ii) evaporation rates can be very high and also energy efficient, (iii) no working gas is required in order to evaporate materials, and (iv) the evaporant flux has relatively low kinetic energy. However, coevaporation of different materials is difficult and evaporation is not consistent with current single-wafer reactor designs in semiconductor production equipment.

Sputter deposition is widely used to deposit metals and other materials in both semiconductor and non-semiconductor applications. The technique has reached a high level of maturity. Many references are available on the subject of sputter deposition (see Rossnagel[64] and references therein). The advantages are simplicity of the equipment, capability of sputtering from targets made of multiple materials, and consistency with current semiconductor equipment design. However, sputter processes are very inefficient from an energy usage standpoint, working gas is generally required in order to sputter, and the flux of sputtered materials has a substantial kinetic energy distribution.

III. Experiments with Evaporated Copper

Holber and coworkers[38] carried out an extensive set of experiments using evaporated copper to examine the capabilities of ECR I-PVD to *fill* high aspect ratio features in multilevel, high-density interconnects. The experimental apparatus is shown in Fig. 1. The experimental test bed consists of a 30.5-cm o.d. stainless-steel vacuum vessel, which is pumped by a turbomolecular pump. Base pressure for the reactor is approximately 5×10^{-7} Torr. Four separate, independently powered electromagnets, having inside diameters of 46 cm, are mounted on rails so that they can be moved along the length of the vacuum vessel. The 125-mm silicon substrates are held by a helium backside-cooled electrostatic chuck. The chuck is equipped with a fiber-optic temperature monitor which can accurately measure the wafer temperature from the backside of the wafer. The wafer chuck allows for the application of either DC or radio frequency (RF) bias to the substrate and is attached to a movable stage with a large range of motion. The system is equipped with a manual single-wafer loadlock which allows the introduction of wafers into the chamber without significant degradation of the chamber base pressure.

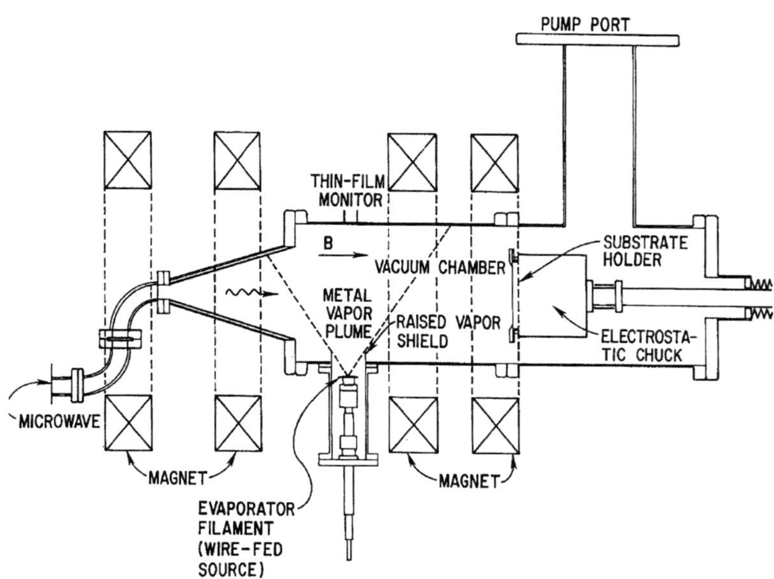

FIG. 1. ECR evaporated copper apparatus (from Holber *et al.*[38]). The copper is introduced into the vacuum chamber from a thermal evaporator located underneath the chamber.

The microwaves are launched into the vacuum chamber as shown in Fig. 1. The microwave/vacuum window (quartz construction) is positioned in the microwave waveguide with a right-angle bend separating the window from the actual deposition chamber. This prevents metal species from the plasma from depositing onto the window, which would block the microwave propagation. It also eliminates thermal loading of the window from plasma bombardment.

The copper evaporator is positioned underneath the vacuum chamber. The copper is fed into a resistively heated assembly (molybdenum or tungsten boat) by a remotely located motor-driven wire feeder. The capacity of the wire feeder allows for a large number of wafer depositions between servicing. The feed rate of the copper wire was set to yield a desired evaporation rate into the vacuum chamber. A typical evaporation rate was 1.0 g/min of copper.

The copper evaporator and wafer chuck were positioned such that with the plasma off, evaporated copper did not reach the substrate. This was verified by turning on the evaporator with the plasma off and examining the wafer for any deposition. The evaporated copper flux also could not reach the quartz microwave/vacuum window.

Located above the vacuum chamber opposite the copper evaporator was a quartz crystal deposition monitor. This monitor samples the evaporated flux and allows measurement of the relative change of flux to it, as conditions in either the evaporator or the plasma are changed.

When the substrate is biased with DC voltage it can be used to measure the ion saturation current at the substrate, giving the arrival rate of copper ions. Using substrate current as a diagnostic (with DC bias voltage set to -50 V) along with the relative measurement of neutral copper atoms reaching the quartz crystal monitor, it can be seen in Fig. 2 that for a given copper evaporation rate as the absorbed microwave power is increased, the ion flux to the substrate increases and the flux to the quartz crystal monitor decreases. This is an indication of depletion of the neutral evaporant feed in the body of the plasma. In additional experiments, as the copper evaporation rate and the microwave power were increased, the ion flux to the substrate continued to increase.

Comparison of the ion flux reaching the substrate and the copper deposition rate confirms that the deposited material is due solely to ions. The key parameter controlling the deposition characteristics into features is the substrate bias, which affects both ion bombardment energy and ion directionality. From work reported in the literature with bias sputter deposition of SiO_2,[65] it is clear that the resputtering rate of the copper ions, as a function of applied voltage, is a critical factor. Figure 3 shows the deposition rate of the copper ions as a function of applied DC voltage at

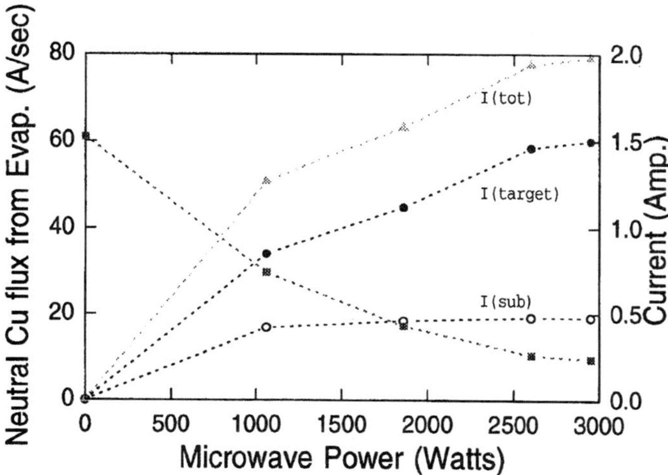

FIG. 2. Measurement of depletion of flux of copper neutral atoms as a function of microwave power. Work was carried out on the apparatus shown in Fig. 1 and is from Holber *et al.*[38] The neutral flux is measured with a quartz-crystal monitor and the current is measured on the 125-mm wafer (I_{sub}) and on the shield surrounding the wafer (I_{target}) to give a total ion flux in the plane of the wafer (I_{tot}).

the substrate. Note that at about −250 V applied DC bias, the curve crosses over from net deposition into net etch.

Numerous deposition experiments were carried out in order to characterize as a function of substrate bias the deposition profiles into high aspect ratio features. The dielectric features to be filled are composed of polyimide with appropriate liner materials. At zero or low applied DC bias voltage, even modest aspect ratio via holes could not be filled completely. Figure 4 shows the results of filling features having an aspect ratio of 1.4 for various deposition times. The applied bias is −50 V. As the features become filled the deposition rate within the feature decreases. This is due to the fact that the resputtered copper is captured within the feature at the start of the process but escapes into the vacuum chamber as the feature becomes more filled. At the −50 V bias voltage the top of the feature pinches off before the feature is filled, resulting in an encapsulated void.

Figure 5 shows the most aggressive features that were successfully filled. This was done using a two-step process, with −230 V bias used for the beginning of the fill and −100 V bias used for the rest of it. Features having aspect ratios of 3 are solidly filled; those having aspect ratios of 4 are filled

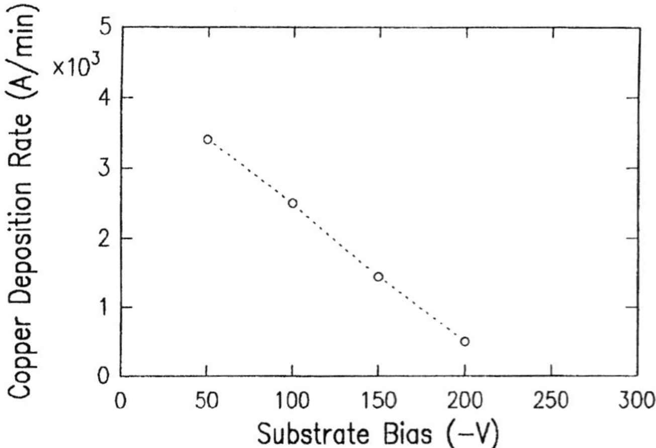

FIG. 3. Resputtering rate of copper ions for the ECR evaporated copper apparatus shown in Fig. 1. Work is from Holber *et al.*[38] The net deposition rate is plotted as a function of applied DC bias voltage.

also, though with some porosity in the grain structure, illuminated by the acid etch used to prepare the samples.

Finally, Fig. 6 demonstrates the utility of the process for application of conformal liners. In this case a single deposition step was used, with an applied −170 V DC bias voltage.

IV. Experiments with Sputtered Materials

The work using evaporated copper[38] demonstrated that useful deposition profiles can be obtained when a flux consisting solely of copper ions is available at the substrate. The controlling factor is the applied substrate bias voltage. However, both in semiconductor and in other materials processing applications, evaporation as a technique has been replaced by sputter deposition because of ease of maintenance, better consistency, and applicability to a wider set of materials, including those with high melting points and alloys and mixtures.

However, the physics both in the plasma and at the substrate is considerably more complicated in the case of sputtered processes. The use of a working gas such as argon is generally required to sustain the sputtered plasma. Atoms sputtered from the target surface are ejected with a broad

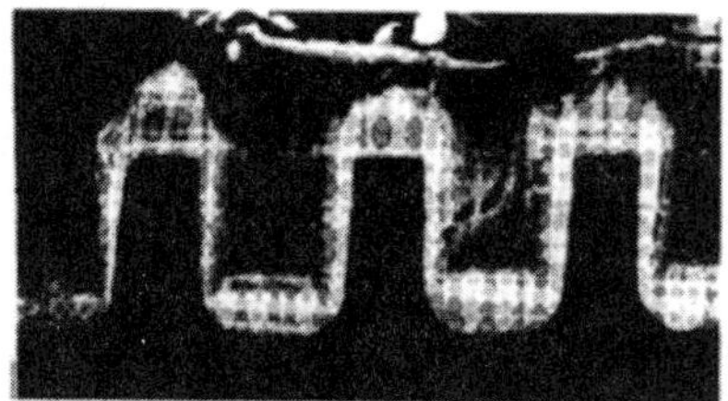

a.

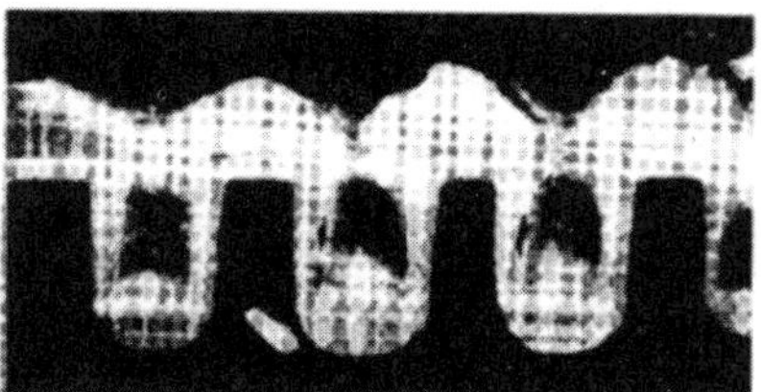

b.

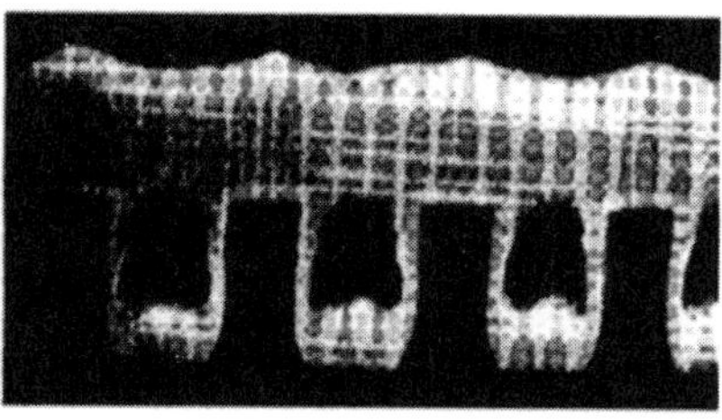

c

FIG. 4. Filling of features having aspect ratio 1.4, using the ECR evaporated copper apparatus shown in Fig. 1. Deposition times are (a) 1 min; (b) 2 min; and (c) 4 min. The applied substrate bias is -50 V DC.

range of kinetic energies, resulting in a corresponding broad range of transit times within the plasma. The result is that the most energetic atoms may not be ionized at all and those which are ionized may have significant velocity components both normal and parallel to the substrate surface. Thus, the species hitting the substrate surface will consist of neutral depositing species of various energies, ionized depositing species of various

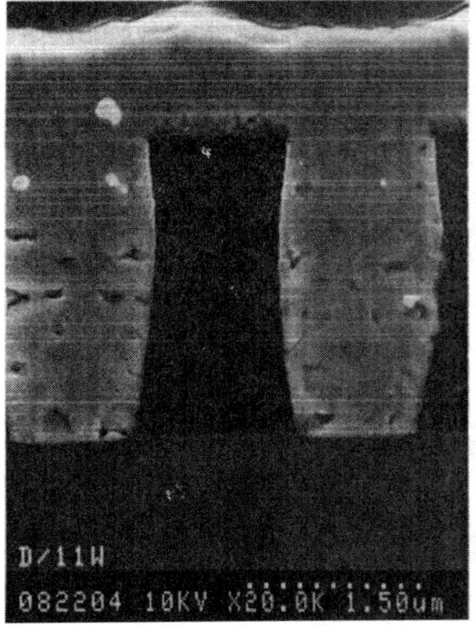

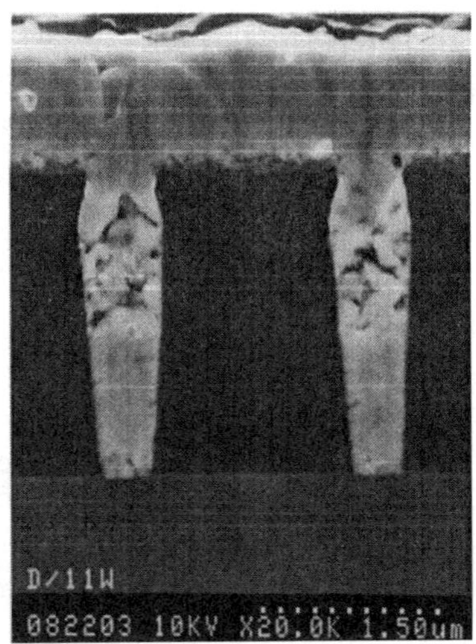

FIG. 5. Deposition of copper into high aspect ratio features using apparatus shown in Fig. 1 (from Holber *et al.*[38]). The process uses two deposition steps, first using −230 V substrate bias, and then decreasing to −100 V-bias.

energies, neutral argon, and ionized argon. There is no convenient way to measure the relative contribution of ions and neutrals to the deposition since the presence of argon ions reduces the usefulness of a simple measurement of current at the substrate surface.

Researchers at Nippon Telegraph and Telephone Corp. (NTT) have carried out numerous studies on ECR sputter deposition.[47–50,54,57] Among the many films deposited are Al_2O_3 and Ta_2O_5;[54] ZnO,[57] Al, Mo, Cu, and

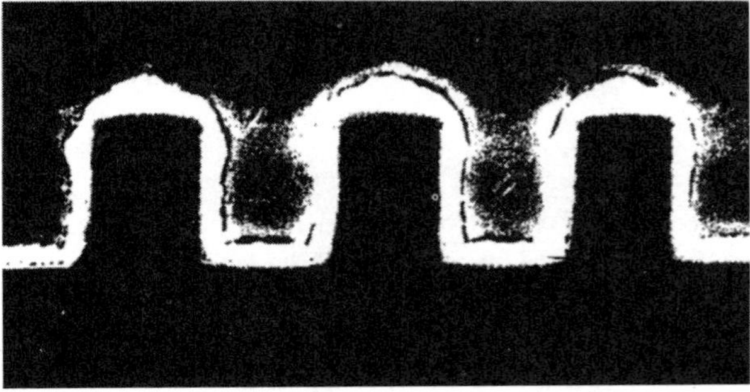

FIG. 6. Deposition of conformal copper liner using ECR evaporated copper apparatus shown in Fig. 1. Substrate bias is set to −170 V (from Holber *et al.*[38]).

Fe[47]; Al, Mo, Fe, and AlN[48]; and Al and TiN.[50] A variety of different microwave launch configurations were studied, with microwave introduction both parallel and perpendicular to the magnetic field. For deposition of non-microwave-absorbing films the quartz microwave/vacuum window could be exposed directly to the plasma.[54]

For the deposition of conductive films a vacuum window that is directly exposed to the plasma cannot be used since it will rapidly become coated with the metal, preventing the microwaves from entering the vacuum chamber. Figure 7 and Fig. 8 show two different approaches to solving this problem. In Fig. 7, the cylindrical quartz window is placed in the microwave combiner structure out of line-of-sight of the plasma. In Fig. 8, the quartz window is placed in the rectangular waveguide, also out of line-of-sight of the plasma.

Kidd[43] used a large ECR plasma system to study the sputtering and deposition of a variety of materials. The reactor was based on a design developed for isotope separation and therefore is of a scale much greater than that of the other systems discussed in this chapter. The vacuum vessel is surrounded by 24 electromagnets. The microwave is at 10.56 GHz, with a corresponding magnetic resonance field strength of 3.8 kG. The microwaves

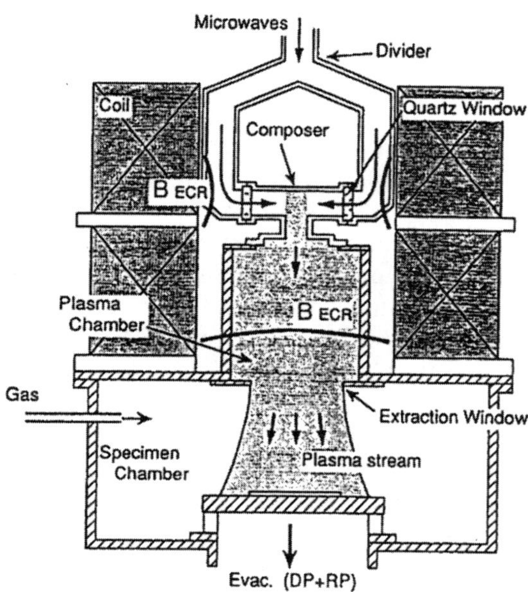

FIG. 7. ECR sputter apparatus from Matsuoka and Ono.[47] The microwave enters the system through a combiner and a quartz window which is out of line of sight of the plasma.

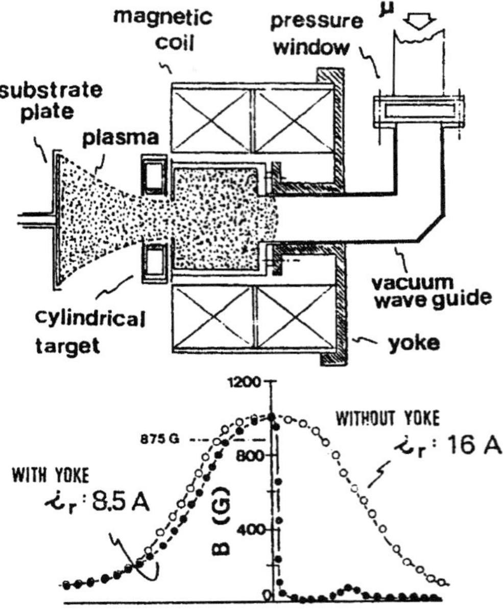

FIG. 8. ECR sputter apparatus from Ono *et al.*[50] The quartz window is inserted in the waveguide, with a right-angle bend separating it from the plasma chamber. The magnetic field strength on axis, centered about the sputter target, is plotted with and without the use of the steel yoke shown.

are introduced into the cylindrical chamber by means of a horn mounted inside the chamber. The sputter target is mounted at the other end of the chamber. The length of the chamber is approximately 150 cm. Although the configuration does not use high-field microwave launch, the high microwave frequency used means that densities in the 10^{12} cm^{-3} range can be obtained.

The ECR plasma is ignited with typically $2–5 \times 10^{-5}$ Torr of argon and the sputter target is then biased at approximately -2 kV. This causes the ejection of atoms into the plasma, which due to the high plasma density and long path length in the plasma are mostly ionized. Measurements indicate that more than 90% of the deposited material on the sample is derived from ions rather than neutrals. For some materials such as copper, the argon gas can be turned off completely during sputter operation, allowing for a pure copper ion plasma. Deposition rates in excess of 1 μm/min were reported.

Horiike and coworkers[45] carried out deposition experiments in an ECR apparatus with Al, Ti, and TiN. The reactor consists of a cryopumped cylindrical vacuum chamber surrounded by multiple electromagnets. The sputter target is coaxial with the vacuum chamber, with the plasma passing through the center of the cylindrically shaped target. The authors found that the plasma characteristics and, correspondingly, the deposition results could be greatly affected by using a magnetic field strength that was twice the normal 875-G resonance value. Deposition rates of 3000 Å/min for aluminum were obtained under conditions in which the estimated fraction of depositing materials which come from ions reached 60%. The filling characteristics of the aluminum were thought to be due to both the high Al ion fraction and surface diffusion.

Musil *et al.*[52,66,67] carried out a study of copper sputtering in an ECR reactor. A cylindrical magnetron was mounted in the central region of the reactor. The microwave input was via a coaxial launch in order to prevent coating of the microwave window with copper. Microwave power was limited to a maximum of 800 W. The film properties of copper deposited onto substrates were studied. By changing argon pressure the energy of the bombarding species was varied, allowing control over the film properties.

Berry and Gorbatkin[32,40–42] describe an ECR reactor which utilizes a permanent magnet instead of an electromagnet in the source region. A microwave launch/window assembly was developed which both allows high-field launch of the microwave and keeps the window out of line of sight of the plasma. The copper sputter target is conical and positioned so that the ECR resonance zone is contained within the conical surface. In addition, the deposition chamber is surrounded by an array of multipole magnets which provide some measure of confinement for the plasma. Plasma densities in the 10^{12} cm^{-3} range were obtained with copper deposition rates in excess of 1000 Å/min.

V. Study of a Highly Ionized ECR I-PVD Reactor

For a project designed to assess the filling capabilities of an ECR-sputtered I-PVD system under conditions of low argon pressure and long ionization path length in the plasma, the experimental apparatus shown in Fig. 9 was developed.[39,68] The deposition chamber is a cylindrical stainless-steel vessel, equipped with a stainless-steel liner. The system is pumped by a 2200 liters/sec turbomolecular pump and has a base pressure of 3×10^{-7} Torr. Typical operating pressure is $0.5–5.0 \times 10^{-3}$ Torr.

The ECR sputter source combines the ECR launch and the sputter target bias in one compact, coaxial assembly. Microwave power at 2.45 GHz is fed into the top of the assembly through a rectangular waveguide. It then passes through a waveguide-to-coax transition and is launched into the vacuum chamber around the 12.5-cm-diameter sputter target. The grounded dark space target shield serves as the inner conductor of the microwave coaxial

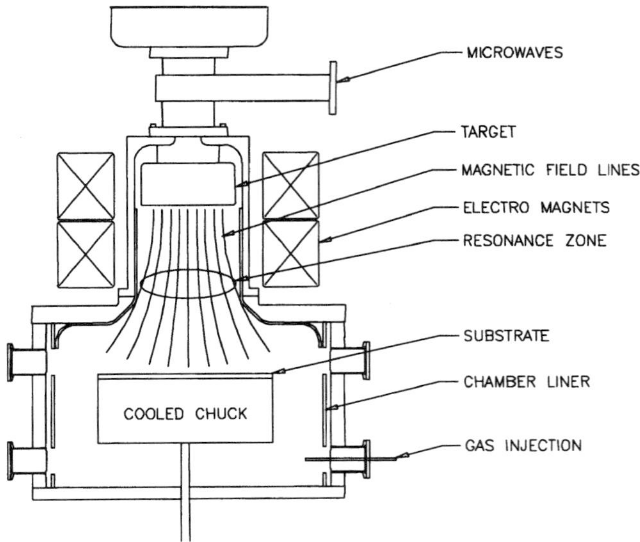

FIG. 9. Diagram of the ECR-sputter apparatus from Chen *et al.*[39] The microwave power is coupled through a coaxial structure around the sputter target and into the vacuum chamber.

transmission line. The vacuum/microwave window consists of a quartz ring that is recessed so that it will not receive direct impingement from the plasma. The quartz window also provides the necessary electrical insulation of the sputter target from the rest of the assembly. The sputter target can be either DC or RF biased by a conductor passing through the center of the coaxial assembly.

Three separately powered external electromagnets, two around the source region and one around the substrate, allow a broad range of magnetic field conditions. The ECR resonance condition is set up in the region between the sputter target and the substrate. The shape and position of the resonance zone and the magnetic field profile can be widely varied.

The substrate holder consists of a water-cooled 8-in.-diameter copper plate. In order to ensure good thermal contact and thus known wafer temperature, the sample wafers could be thermally bonded to the copper plate with a high thermal conductivity vacuum grease. The substrate holder can also be biased in a flexible manner, permitting both DC bias and RF bias of various frequencies.

For the process regime of interest in this reactor, the microwave plasma characteristics are largely independent of either the applied sputter target bias or the applied substrate bias. This allows for independent control of many parameters that are coupled together in conventional sputter deposition systems. In this system,

- Sputter target potential is set by the sputter bias power supply
- Ion current to the sputter target depends on the ECR plasma characteristics and is largely independent of sputter target voltage
- Plasma density is largely independent of sputter target and substrate potentials, and it is determined by microwave power, operating pressure, and magnetic field configuration.
- The fraction of the material depositing onto the substrate which is due to ions is dependent on the plasma density, the electron temperature, and the path length which sputtered species must pass through the plasma.
- The substrate potential is set by the substrate bias power supply

Typical operating parameters for the system are summarized in Table 1.

In order to characterize the basic operation of the reactor, a Langmuir probe was used to measure the plasma properties in the bulk and the substrate and target currents were used to measure the ion flux at the top and bottom edges of the plasma. The critical parameters in determining whether a neutral atom ejected from the sputter target will be ionized are plasma density, path length in the plasma, and electron temperature of the plasma. Even a small change in electron temperature can greatly change the ionization probability. Figure 10 shows a representative graph of electron temperature versus argon pressure; the target voltage is set to $-800\ V_{DC}$. Note the rapid increase in electron temperature with decreasing argon pressure. This is due to the reduced collisionality of electrons with the background argon. On the other hand, the plasma density generally decreases over this same range of argon pressures, as shown by the decrease in target current shown in Fig. 11.

TABLE 1

TYPICAL OPERATING PARAMETERS FOR THE SYSTEM SHOWN IN FIG. 9

Base pressure	2×10^{-7} Torr
Operating pressure	$0.5–5.0 \times 10^{-3}$ Torr
Argon flow rate	20–200 sccm
Target-to-substrate distance	25–38 cm
Microwave power	1.0–5.0 kW
Sputter target DC bias voltage	0–1800 V
Sputter target current	0–8 Å

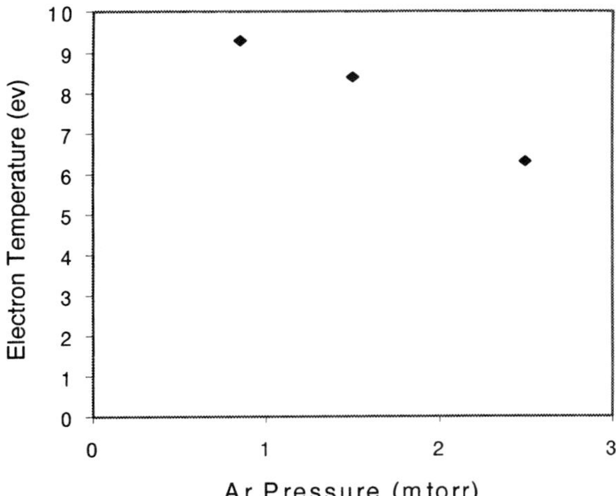

FIG. 10. Electron temperature as a function of argon pressure for the ECR sputter apparatus shown in Fig. 9. The measurement is made with a Langmuir probe. The microwave power is 4.0 kW and the target bias is −800 V.

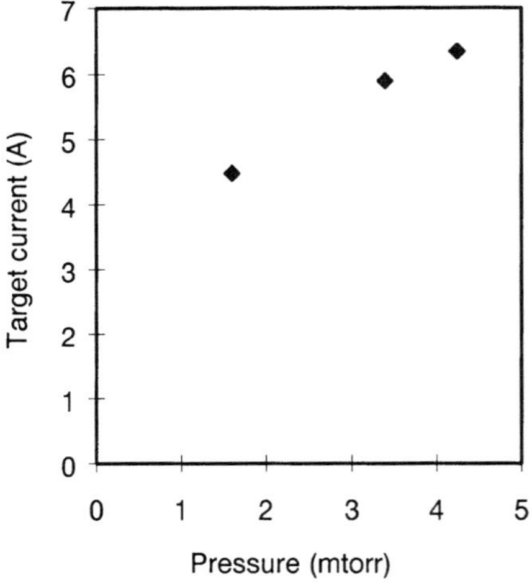

FIG. 11. Sputter target current as a function of argon pressure for the ECR sputter apparatus shown in Fig. 9. The microwave power is 4.0 kW. The sputter target is aluminum and the target voltage is −1000 V.

The effect of microwave power on sputter target current is shown in Fig. 12. The current increases nearly linearly with microwave power, indicating that the microwave coupling into the plasma remains smooth and linear in this magnetic field configuration.

As the sputter target bias voltage is increased the sputter rate will also increase. In work with inductively coupled I-PVD plasmas,[69–72] it has been reported that as more metal from the sputter target is introduced into the plasma, the target current decreases. The possible causes include lowering of the electron temperature and heating of the background argon gas. Figure 13 shows sputter target current as a function of sputter target voltage. Note that while there is a slight decline in target current with increased voltage, it is a very weak dependence. The combination of the high (4-kW) microwave power and the low (1.6×10^{-3} Torr) background pressure appears to put the ECR-PVD system in a different operating regime relative to the referenced inductively coupled systems, which typically operate at lower RF powers and much higher background argon pressures.

The basic operation of the system was studied by carrying out blanket depositions on silicon wafers. The deposition rate on the silicon substrate will be dependent on both the rate at which material is sputtered from the

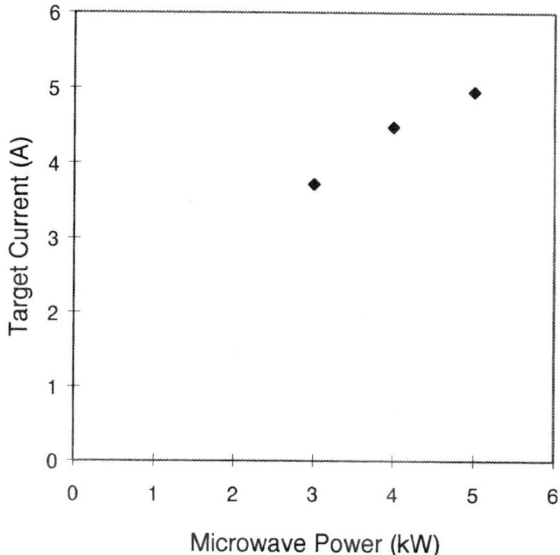

FIG. 12. Sputter target current is shown as a function of microwave power for the ECR sputter apparatus shown in Fig. 9. The sputter target is aluminum and the target voltage is −1000 V. The argon pressure is 1.6 mTorr.

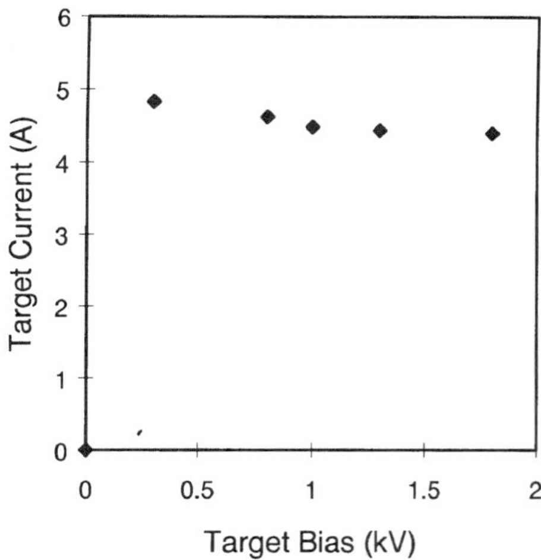

FIG. 13. Sputter target current as a function of target bias voltage for the ECR sputter apparatus shown in Fig. 9. The microwave power is 4.0 kW and the argon pressure is 1.6 mTorr.

target and the efficiency with which the sputtered material is transported to the substrate. Figures 14–16 show the deposition rate as a function of microwave power, argon pressure, and target voltage, respectively. Note the linear dependence of deposition rate on microwave power. The deposition rate increases with sputter target voltage, although at the highest target voltages the increase begins to level off, which is consistent with the sputter behavior of most materials. The behavior with argon pressure is more complex. While Fig. 11 shows that target current increases with pressure, Fig. 10 shows a decrease of electron temperature with increased pressure. Increased argon pressure will therefore increase the sputter rate of material into the plasma but will decrease the transport of material to the substrate through both increased scattering of the sputtered material from background argon and possibly through decreased ionization of the sputtered material due to the lowered electron temperature.

The probability of a sputtered atom being ionized in the plasma is dependent on the translational energy of the atom, the plasma density, the path length in the plasma, and the electron temperature.[73–75] The energy distribution of sputtered species has a characteristic functional form[76–81] in which the most probable energy is typically a few electron volts. However, the long high-energy tail indicates that the average is much higher. Figure 17 shows a calculated energy distribution for sputtered Al. Note that a substantial number of the sputtered atoms have energy >20 eV. Figure 18

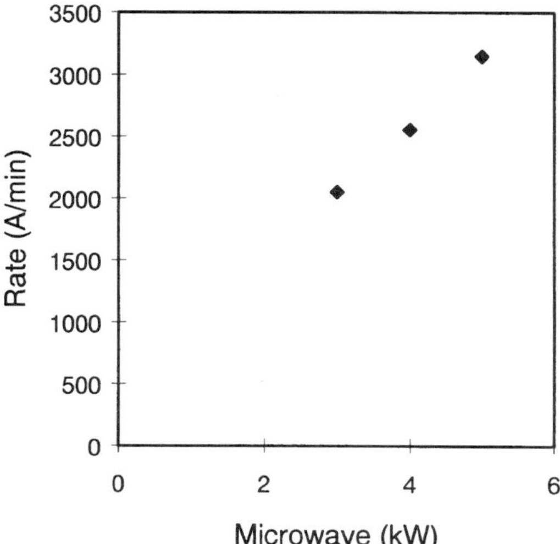

FIG. 14. Blanket deposition rate of aluminum as a function of microwave power for the ECR sputter apparatus shown in Fig. 9. The target bias is −1000 V and the argon pressure is 1.6 mTorr.

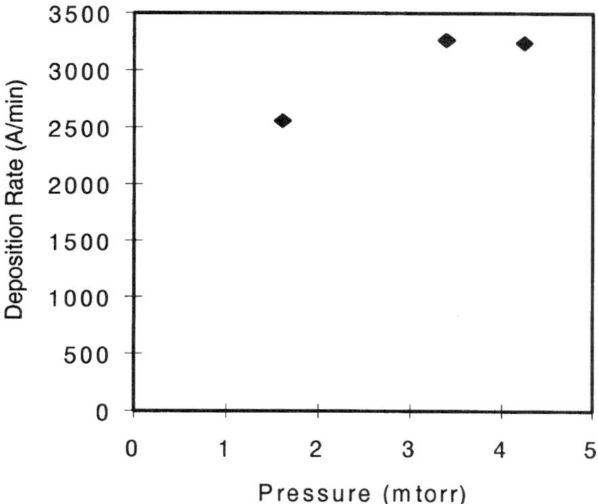

FIG. 15. Blanket deposition rate of aluminum as a function of argon pressure for the ECR sputter apparatus shown in Fig. 9. The microwave power is 4.0 kW and the target voltage is −1000 V.

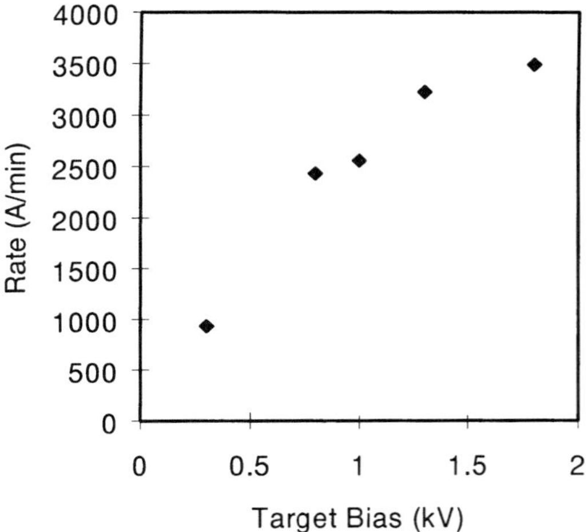

FIG. 16. Blanket deposition rate of aluminum as a function of target voltage for the ECR sputter apparatus shown in Fig. 9. The microwave power is 4.0 kW and the argon pressure is 1.6 mTorr.

shows the calculated ionization path length for Al atoms in a typical plasma in the ECR apparatus as a function of Al atom energy.

Another important factor is the degree to which the sputtered atoms are thermalized by the background argon. While at higher argon pressures the thermalization can be nearly complete,[68–70,82,83] in the ECR pressure

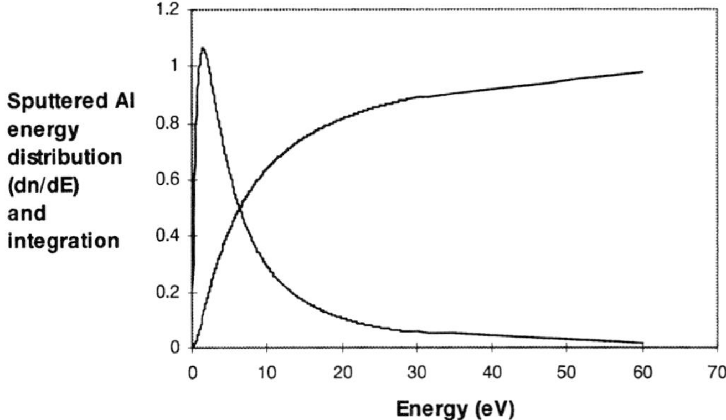

FIG. 17. Calculated energy distribution for sputtered aluminum atoms. No thermalization in background argon is assumed. Both the differential and the integrated calculations are shown. Note that a significant number of sputtered atoms have energies >20 eV.

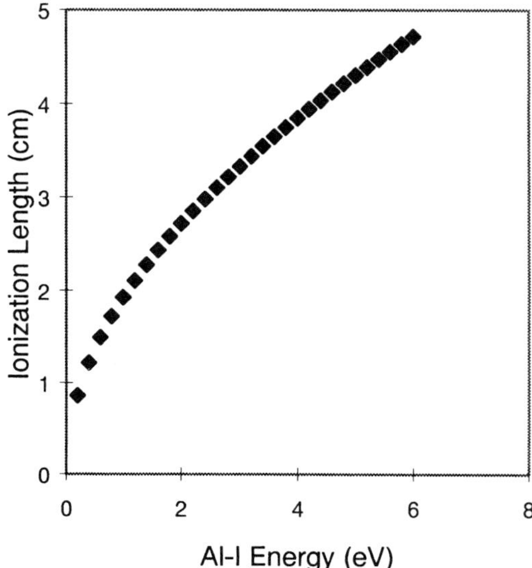

FIG. 18. Calculated ionization path length in the plasma for sputtered aluminum atoms. The assumed plasma density and electron temperature for the calculation are $2 \times 10^{12}\,cm^{-3}$ and 3 eV, respectively.

regime the results are not obvious. A detailed calculation was not carried out to determine this precisely. Nevertheless, from Figs. 17 and 18 and from a comparison of the actual deposition rate versus the expected deposition rate, we estimate that between 50 and 100% of the depositing species are ions rather than neutrals.

Numerous deposition experiments were carried out with both aluminum and copper to assess the capabilities in filling high aspect ratio features. The results obtained in the ECR–PVD reactor can be summarized as follows:

- Substrate bias is the most important variable in fill capability, for both copper and aluminum.
- The effect of other process parameters is primarily on rate and only secondarily on fill.
- The highest aspect ratio fills achieved for Cu were approximately 1.9.
- The highest aspect ratio fills achieved for Al were approximately 3.0, with no upper limit seen.
- The difference between Al and Cu fill characteristics has to do with processes at the substrate and not in the plasma (i.e., resputter rate, sputter angle, sticking coefficient, and surface mobility).

Figure 19 shows SEMs of trenches deposited with Al, with no applied substrate bias voltage. Trenches of aspect ratio 2.0 or less can be filled in this manner. Simulations indicate that these results are consistent with ion temperature of approximately 0.2 eV, with 10-V sheath potential.

Figure 20 shows SEMs of trenches deposited with Al, with DC bias applied to the substrate. The best fill results are obtained with an applied bias between −30 and −40 V. Above −40 V the etching at the substrate increases faster than the deposition, causing damage to the features. Below −30 V the resputtering is not fast enough to keep the buildup off the corners of the features, which is necessary in order to avoid pinch-off of the feature.

While the basic trends for copper are similar to those for aluminum, the fill results obtained are not as aggressive. Figure 21 shows features deposited with Cu under conditions of no applied substrate bias. Note that the aspect ratio 1.9 features are not filled. Figure 22 shows the results with DC bias. Although filling improves with bias voltage, at −35 V, which is the highest

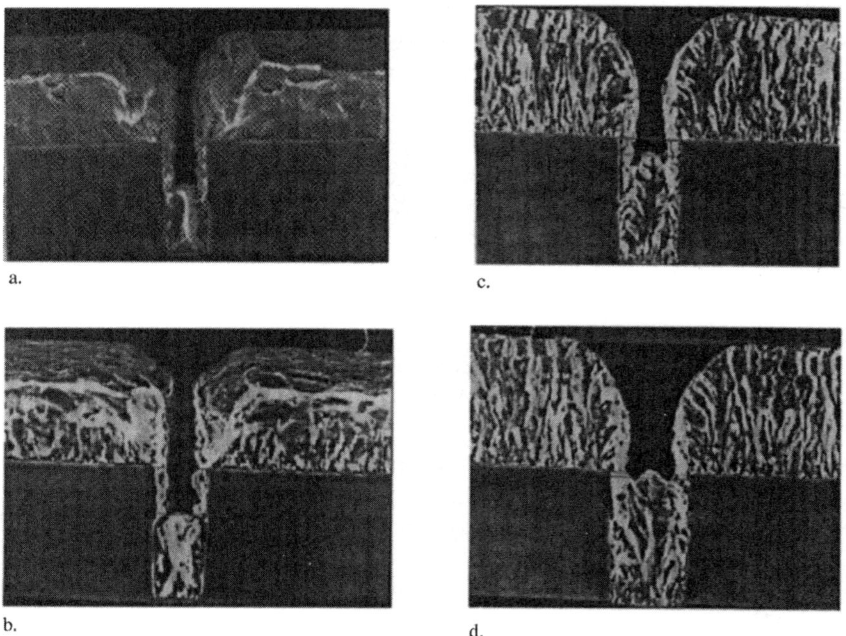

FIG. 19. Trench features with aluminum deposited under condition of no applied substrate bias from the ECR sputter apparatus shown in Fig. 9. The depth of the trenches is 1.2 μm. The widths are (a) 0.4 μm, (b) 0.5 μm, (c) 0.6 μm, and (d) 0.7 μm. The microwave power is 4.0 kW, the target voltage is −1000 V, and the argon pressure is 1.6 mTorr.

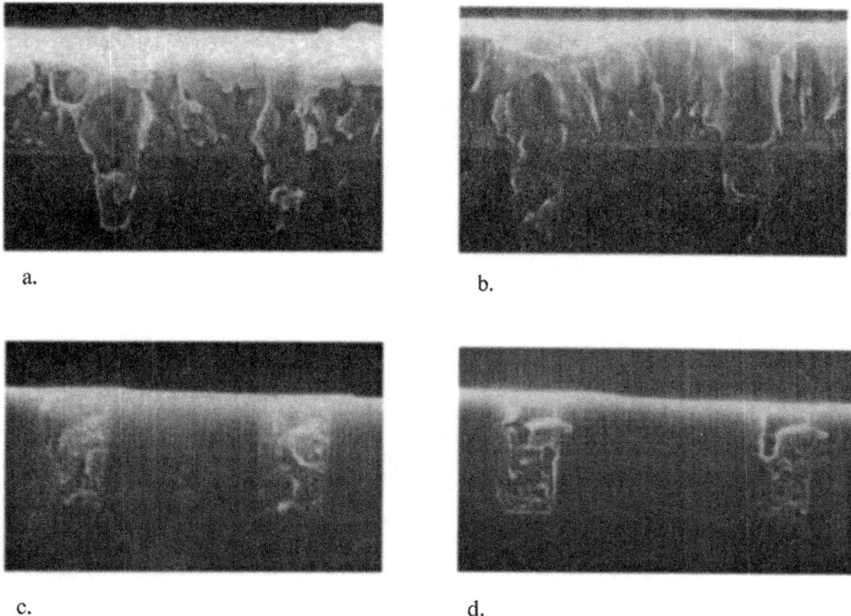

FIG. 20. Features filled with Al under various DC bias conditions from the ECR sputter apparatus shown in Fig. 9. The depth of the features is 1.2 μm in all cases. The DC bias conditions and feature widths are (a) -10 V, 0.55 μm; (b) -20 V, 0.55 μm; (c) -30 V, 0.65 μm; and (d) -40 V, 0.65 μm. The sputter target voltage is -600 V, the microwave power is 4.0 kW, and the argon pressure is 1.5 mTorr. Note that the best fill is achieved between -30 and -40 V bias.

voltage that can be used without damaging the feature, higher aspect ratio features still cannot be filled. Figure 23 show the most aggressive Cu fills achieved. A multistep process is used in which the first part of the fill is carried out using -30-V applied bias, followed by a zero-bias step. There is still a slight void in the aspect ratio 2.0 features.

In summary, substrate bias dominates the fill characteristic in the process regime investigated in this reactor. It is expected that the majority of the depositing species are composed of ions in all the cases studied. Features with aspect ratios of 3 were filled with aluminum, with no limit seen when increasing aspect ratios. For copper the situation is different and fills of aspect ratios more than about 1.9 were not achieved. Finally, in addition to the copper and aluminum fill studies, the same apparatus has been used to successfully deposit conformal liners of materials such as Ti, TiN, Ta, and Cu.

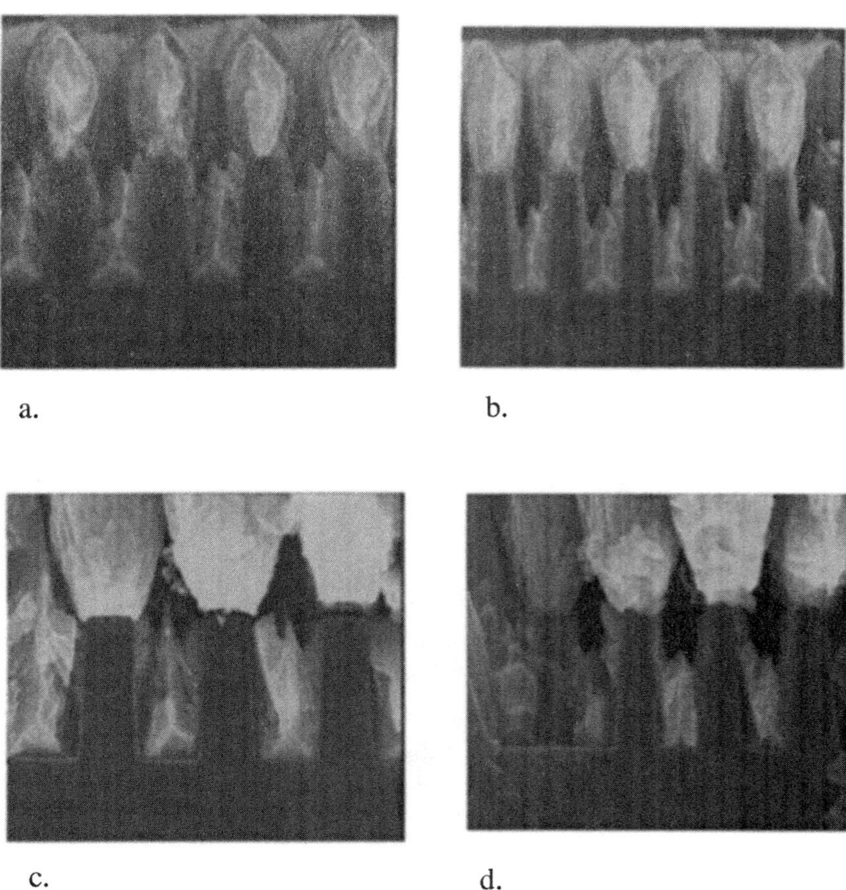

FIG. 21. Trench structures with copper deposited under conditions of no applied substrate bias from the ECR sputter apparatus shown in Fig. 9. The trench depth is 1.2 μm in all cases. The microwave power is 2.5 kW in a and b and 5.0 kW in c and d. The target voltage and argon pressure in all the cases are −1200 V and 2.5 mTorr, respectively. The trench width and corresponding aspect ratio (AR) are (a) 0.65 μm, AR 1.9; (b) 0.5 μm, AR 2.4; (c) 0.65 μm, AR 1.9; (d) 0.5 μm, AR 2.4.

VI. Conclusions

ECR reactors have been used to carry out ionized PVD depositions in many reported studies. The applications range from metallization for semiconductor applications to hard coatings. While studies using evaporative sources

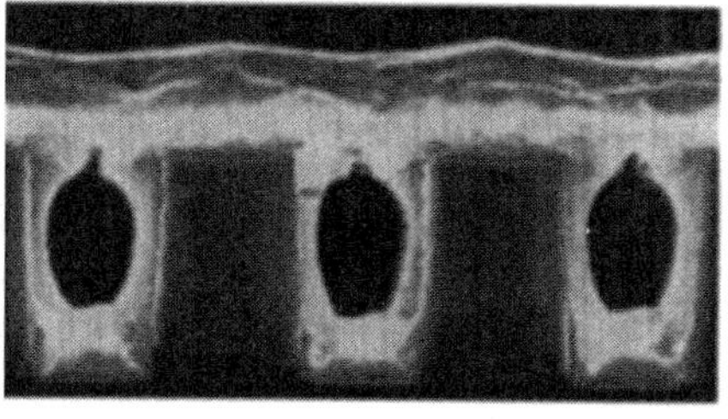

a.

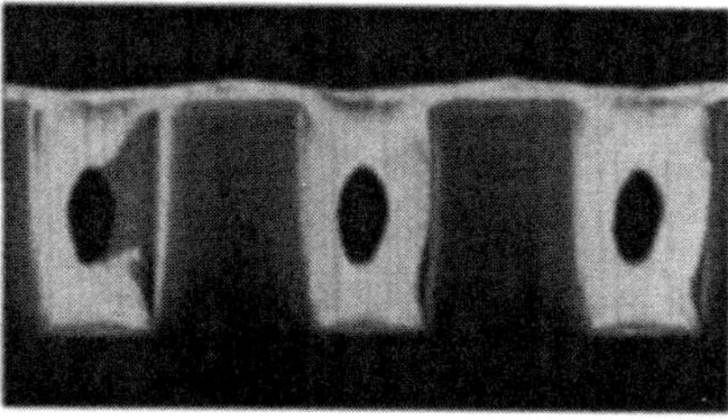

b.

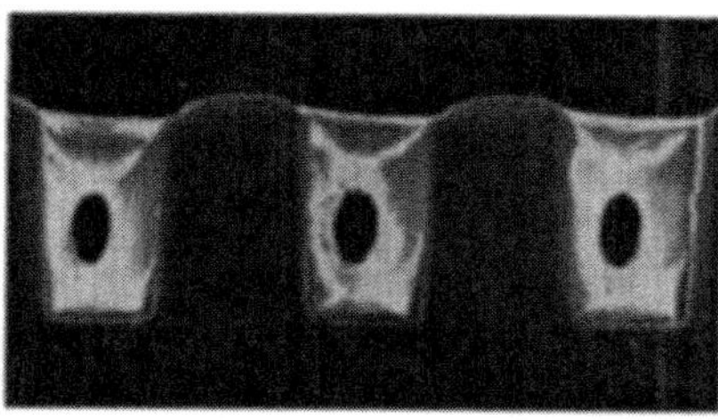

c.

FIG. 22. Trench structures with copper deposited under conditions of DC substrate bias from the ECR sputter apparatus shown in Fig. 9. The trench depth is 1.2 μm in all cases. The microwave power is 4.0 kW, the argon pressure is 1.5 mTorr, and the target voltage is −800 V. The applied substrate bias is (a) −25 V, (b) −30 V, and (c) −35 V.

allow basic physical mechanisms to be elucidated, most of the applications have been carried out using sputter-based sources.

It is important to consider where ECR I-PVD sources may play an important role relative to other means of vacuum deposition. The advantages of ECR I-PVD are strongest when some or all the following are desired: (i) very low background pressures during the deposition, (ii) the highest level of ionization of the depositing material, (iii) independent control over the flux and energy of the depositing material, and (iv)

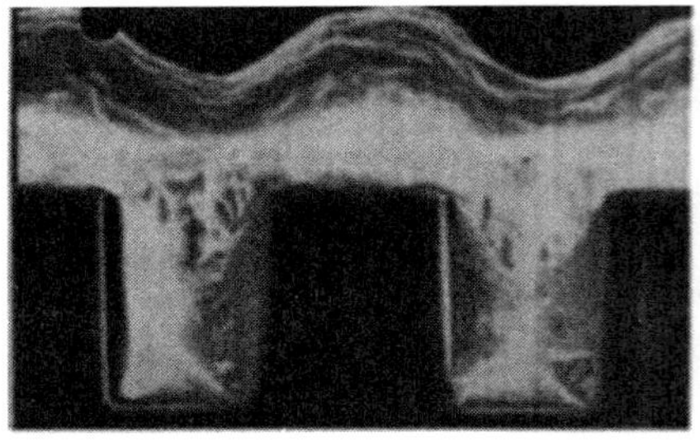

a.

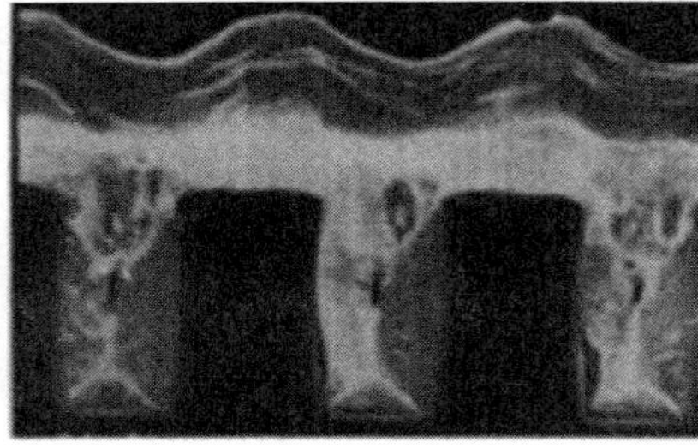

b.

FIG. 23. Trench structures with copper deposited under two-step DC substrate bias from the ECR sputter apparatus shown in Fig. 9. The trench depth is 1.2 μm in both cases. The trench width and corresponding aspect ratio are 0.8 μm and AR 1.5 for a and 0.6 μm and AR 2.0 for b. The microwave power is 4.0 kW, the argon pressure is 1.5 mTorr, and the target voltage is −800 V. The wafer bias is −30 V followed by a period with no applied bias.

independent control of the sputter target voltage and current. In such circumstances, ECR I-PVD can provide a unique tool for thin film deposition, allowing deposition over difficult features and allowing unique film properties in a practical deposition manner.

Acknowledgments

I would like to acknowledge the help of the many coworkers who assisted with the experiments in ECR deposition using both evaporated and sputtered sources. In addition, Jeff Hopwood assisted in carrying out the ionization path length calculations.

References

1. A. C. England, *IEEE Trans. Plasma Sci.* **PS-12**, 124 (1984).
2. K. Suzuki, S. Okudaira, N. Sakudo, and I. Kanamoto, *Jpn. J. Appl. Phys.* **16**, 1979 (1977).
3. N. Sakudo, K. Tokiguchi, H. Koike, and I. Kanomata, *Rev. Sci. Instrum.* **48**, 762 (1977).
4. N. Sakudo, K. Tokiguchi, H. Koike, and I. Kanomata, *Rev. Sci. Instrum.* **49**, 940 (1978).
5. M. A. Heald and C. B. Wharton, *Plasma Diagnostics with Microwaves*, (Krieger, Melbourne, FL, 1978.)
6. F. F. Chen, *Introduction to Plasma Physics and Controlled Fusion*, Plenum, New York, 1984.
7. T. H. Stix, *The Theory of Plasma Waves*, McGraw-Hill, New York, 1962.
8. J. S. McKillop, J. C. Forster, and W. M. Holber, *J. Vac. Sci. Technol. A* **7**, 908 (1989).

9. G. Guan, M. E. Mauel, W. M. Holber, and J. B. O. Caughman, *Phys. Fluids B* **4**, 4177 (1992).
10. J. Musil, *Vacuum* **36**, 161 (1986).
11. J. Asmussen, T. A. Grotjohn, P. Mak, and M. A. Perrin, *IEEE Trans. Plasma Sci.* **25**, 1196 (1997).
12. T. Nakano, N. Sadeghi and R. A. Gottscho, *Appl. Phys. Lett.* **58**, 458 (1991).
13. Y. Torii, M. Shimada, and I. Watanabe, *Nucl. Instrum. Methods Phys. Res. B* **21**, 178 (1987).
14. E. A. Den Hartog, H. Persing, and R. C. Woods, *Appl. Phys. Lett.* **57**, 661 (1990).
15. T. D. Mantei and T. Wicker, *Appl. Phys. Lett.* **43**, 84 (1983).
16. T. D. Mantei and T. E. Ryle, *J. Vac. Sci. Technol. B* **9**, 29 (1991).
17. K. Kato and I. Kato, *Jpn. J. Appl. Phys.* **28**, L343 (1989).
18. H. Fujita, H. Handa, M. Nagano, and H. Matsuo, *Jpn. J. Appl. Phys.* **26**, 1112 (1987).
19. J. M. Essick, F. S. Pool, Y. H. Shing, and M. J. Holboke, Proceedings of the Materials Research Meeting, April 1991.
20. J. C. Rostaing, F. Coeuret, J. Pelletier, T. Lagarde, and R. Etemadi, *Thin Solid Films* **270**, 49 (1995).
21. H. Yamada and Y. Torii, *J. Appl. Phys.* **65**, 1106 (1989).
22. K.-C. Wang, K.-L. Cheng, Y.-L. Jiang, T.-R. Yew, and H.-L. Hwang, *Jpn. J. Appl. Phys.* **34**, 927 (1995).
23. B. Lane, S. Lossig, W. Harshborger, and W. Holber, Proceedings of Symposium on Dry Process, Nov. 1–3, 1995, Tokyo, Japan, p. 287.
24. K. Machida, C. Hashimoto and H. Oikawa, *J. Vac. Sci. Technol. B* **11**, 224 (1993).
25. O. A. Popov and H. Waldron, *J. Vac. Sci. Technol. A* **7**, 914 (1989).
26. C. S. Pai, J. F. Miner, and P. D. Foo, *J. Electrochem. Soc.* **139**, 850 (1992).
27. M. Kitagawa, T. Hirao, T. Ohmura and T. Izumi, *Jpn. J. Appl. Phys.* **6**, L1048 (1989).
28. C. Keqiang, Z. Erli, W. Jinfa, Z. Hansheng, G. Zuoyao, and Z. Bangwei, *J. Vac. Sci. Technol. A* **4**, 828 (1986).
29. S. Matsuo and M. Kiuchi, *Jpn. J. Appl. Phys.* **22**, L210 (1983).
30. T. V. Herak, T. T. Chau, D. J. Thompson, S. R. Mejia, D. A. Buchanan and K. C. Kao, *J. Appl. Phys.* **65**, 2457 (1989).
31. T. Fukuda, M. Ohue, N. Momma, K. Suzuki and T. Sonobe, *Jpn. J. Appl. Phys.* **6**, 1035 (1989).
32. S. M. Gorbatkin, R. L. Rhoades, T. Y. Tsui and W. C. Oliver, *Appl. Phys. Lett* **65**, 1672 (1994).
33. T. Maeda, H. Nakae, and T. Hirai, ISPC-8, p. 2434, Tokyo, 1987.
34. A. Chayahara, H. Yokoyama, T. Imura, Y. Osaka and M. Fujisawa, ISPC-8, p. 2440, Tokyo, 1987.
35. T. Akahori, A. Tanihara and M. Tano, *Jpn. J. Appl. Phys.* **30**, 3558 (1991).
36. K. Endo, T. Miyamura, N. Kitaroi *et al.*, *Jpn. J. Appl. Phys.* **37**, 3486 (1998).
37. S. O. Chung, J. W. Kim, S. T. Kim *et al.*, *Mater. Chem. Phys.* **53**, 60 (1998).
38. W. M. Holber, J. S. Logan, H. J. Grabarz, J. T. C. Yeh, J. B. O. Caughman, A. Sugerman, and F. E. Turene, *J. Vac. Sci. Technol. A* **11**, 2903 (1993).
39. W. Holber, X. Chen, L. Bourget, J. Urbahn, S. Jin, T. Y. Yao, K. Ngan, Z. Xu, and S. Ramaswami, submitted for publication.
40. L. A. Berry, S. M. Gorbatkin, and R. L. Rhoades, *Thin Solid Films* **253**, 382 (1994).
41. C. Doughty, S. M. Gorbatkin, and L. A. Berry, *J. Appl. Phys.* **82**, 1868 (1997).
42. S. M. Gorbatkin, D. B. Poker, R. L. Rhoades, C. Doughty, L. A. Berry, and S. M. Rossnagel, *J. Vac. Sci. Technol. B* **13**, 1853 (1996).
43. P. Kidd, *J. Vac. Sci. Technol. A* **9**, 466 (1991).

44. T. Asamaki, T. Miura, A. Takagi, R. Mori, and K. Hirata, *Jpn. J. Appl. Phys.* **33**, 4566 (1994).
45. S. Shingubara, N. Morimoto, S. Takehiro, Y. Matsui, I. Utsunomiya, Y. Horiike, and H. Shindo, *Appl. Phys. Lett.* **63**, 737 (1993).
46. S. Shibuki, H. Kanao, and T. Akahori, *J. Vac. Sci. Technol. B* **15**, 60 (1997).
47. M. Matsuoka and K. Ono, *Jpn. J. Appl. Phys.* **28**, L503 (1989).
48. M. Matsuoka and K. Ono, *Appl. Phys. Lett.* **54**, 1645 (1989).
49. M. Matsuoka, *Mater. Sci. Forum* **140–142** , 55–78 (1993).
50. T. Ono, H. Nishimura, M. Shimada, and S. Matsuo, *J. Vac. Sci. Technol. A* **12**, 1281 (1994).
51. Y. Yoshida, *Appl. Phys. Lett.* **61**, 1733 (1992).
52. J. Musil and M. Misina, *Czech. J. Phys.* **46**, 353 (1996).
53. P. Mueller, W. M. Holber, W. Henrion, E. Nebauer, V. Schlosser, B. Selle, I. Sieber, and W. Fuhs, Low-temperature deposition of microcrystalline silicon by microwave plasma-enhanced sputtering, in *Polycrystalline Semiconductors*, (J. H. Werner, H. P. Strunk, and H. W. Schock, Eds.), Schwäbisch Gmünd, Germany, 1998.
54. C. Takahashi, M. Kiuchi, T. Ono, and S. Matsuo, *J. Vac. Sci. Technol.* **6**, 2348 (1988).
55. J. C. Barbour, D. M. Follstaedt, and S. M. Myers, *Nucl. Inst. Methods Phys. Res. B* **106**, 84 (1995).
56. M. Delaunay and E. Touchais, *Rev. Sci. Inst.* **69**, 2320 (1998).
57. M. Matsuoka and K. Ono, *Appl. Phys. Lett.* **53**, 1393 (1988).
58. S. Takebayashi and K. Shimokawa, *J. Appl. Phys.* **69**, 5673 (1991).
59. H. Masumoto, T. Goto, Y. Masuda, A. Baba, and T. Harai, *Appl. Phys. Lett.* **58**, 243 (1991).
60. M. Misina, Y. Setsuhara, and S. Miyake, *Jpn. J. Appl. Phys.* **36**, 3629 (1997).
61. M. Misina, Y. Setsuhara, and S. Miyake, *J. Vac. Sci. Technol. A* **15**, 1922 (1997).
62. G. M. Rao and S. B. Krupanidhi, *Appl. Phys. Lett.* **70**, 628 (1997).
63. T. Aida, A. Tsukamoto, K. Imagawa, T. Fukazawa, S. Saito, K. Shindo, K. Takagi, and K. Miyauchi, *Jpn. J. Appl. Phys.* **28**, L635 (1989).
64. S. M. Rossnagel, *J. Vac. Sci. Technol. B* **16**, 2584 (1998).
65. J. S. Logan, M. J. Hait, H. C. Jones, G. R. Firth, and D. B. Thompson, *J. Vac. Sci. Technol. A* **7**, 1392 (1989).
66. J. Musil, M. Misina, and D. Hovorka, *J. Vac. Sci. Technol. A* **15**, 1999 (1997).
67. J. Musil, *Vacuum* **47**, 145 (1996).
68. U.S. Patent No. 05688382, issued November 18, 1997.
69. J. Hopwood, *Phys. Plasmas* **5**, 1624 (1998).
70. M. Dickson, F. Qian, J. Hopwood, *J. Vac. Sci. Technol. A* **15**, 340 (1997).
71. S. M. Rossnagel and J. Hopwood, *J. Vac. Sci. Technol. B* **12**, 449 (1994).
72. J. Hopwood and F. Qian, *J. Appl. Phys.* **78**, 758 (1995).
73. W. Lotz, *Astrophys. J.* **14**, 207 (1967).
74. W. Lotz, *Z. Physik* **232**, 101 (1970).
75. L. L. Shimon, E. I. Nepiipov, and I. P. Zapesochnyi, *Sov. Phys. Tech. Phys.* **20**, 434 (1975).
76. G. Betz and K. Wien, *Int. J. Mass Spec. Ion Proc.* **140**, 1 (1994).
77. H. L. Bay, *Nucl. Inst. Methods Phys. Res.* **B18**, 430 (1987).
78. A. Wuchner and H. Oechsner, *Nucl. Inst. Methods Phys. Res.* **B18**, 458 (1987).
79. W. Eckstein, *Nucl. Inst. Methods Phys. Res.* **B18**, 344 (1987).
80. J. Dembowski, H. Oechsner, Y. Yamamura, and M. Urbassek, *Nucl. Inst. Methods Phys. Res.* **B18**, 464 (1987).
81. A. Wuchner and W. Reuter, *J. Vac. Sci. Technol. A* **6**, 2316 (1988).
82. Y. Yamamura and M. Ishida, *J. Vac. Sci. Technol. A* **13**, 101 (1995).
83. G. M. Turner, I. S. Falconer, B. W. James, and D. R. McKenzie, *J. Vac. Sci. Technol. A* **11**, 2758 (1993).

Ionized Hollow Cathode Magnetron Sputtering

KWOK F. LAI

Novellus Systems, San Jose, CA

I. Introduction

The Hollow Cathode Magnetron (HCM) is a new type of high-density plasma device developed for ionized physical vapor deposition (I-PVD). Unlike other I-PVD approaches in which post-ionization of sputtered or evaporated metal atoms by either radiofrequency (RF) or microwave-generated high density plasma is necessary, the HCM uses only a single DC power supply to both sputter and ionize the target material. A novel magnetic geometry provides the confining magnetic field to sustain a magnetron discharge within a cup-shaped hollow cathode and the means of ion extraction to allow the metal plasma to stream to the substrate.

In conjunction with efficient water cooling, the HCM is capable of operating at more than 10 times the power density of conventional planar magnetrons. High power density, together with efficient confinement allow the HCM to achieve extremely high plasma density ($\sim 10^{13}\,\mathrm{cm}^{-3}$). While

Vol. 27
ISBN 0-12-533027-8

ISSN 1079-4050/00 $30.00

I-PVD using RF inductively coupled plasma (RFiPVD) has a plasma density of 10^{11} cm^{-3} to 10^{12} cm^{-3} and operates best above tens of mTorr,[1] the HCM achieves high degrees of ionization at only a few mTorr, primarily due to its extremely high plasma density.

II. Principles of Operation

A. Invention Background

Ion plating[2] is a well-known technique to overcome the "line of sight" and adhesion problems of evaporation. By applying a high negative potential on the substrate to attract positive ions of the plating material, conformal coating can be achieved on substrates of complex geometry. Due to its low deposition rate and high bias voltage, conventional ion plating is not of commercial interest to the semiconductor metallization. The use of ion plating for very large scale integrated circuit (VLSI) multilevel metallization was first reported by Mei *et al.*[3] using partially ionized beam (PIB) deposition. Because of its low ionization ratio ($<5\%$), PIB deposition requires very high acceleration voltage (few kilovolts) to realize the benefit of metal ion deposition. Hollow cathode discharge (HCD)[4] and the pressure gradient-type plasma gun[5] have also been used to enhance ion plating. By operating the HCD or the plasma gun in the low-voltage high-current mode and applying an electric field between the evaporation source and the substrate, metal deposition with high ion to atom ratio and rates up to 1 μm/min for aluminum[6] and 0.4 μm/min for copper[7] can be collected at the substrate. A steering coil and a magnetic field in the proximity of the hearth (plasma beam controller) can be further used to deflect and guide the plasma beam to the hearth. Because of the much higher density plasma ($10^{10}-10^{12}$ cm^{-3}) in the HCD and the plasma gun, the fraction of the metal ions in these enhanced ion plating plasma is much higher than conventional ion plating. However, due to their small size, poor uniformity, and the inherent limitations of evaporation,[8] the HCD and the pressure gradient-type plasma gun are not likely to be an attractive technology for VLSI metalization.

To avoid the use of high bias voltage, to achieve a high degree of ionization in the metal atoms, and to sustain a high deposition rate, a uniform metal plasma of sufficiently high density is required. The development of such a high-density plasma source to efficiently ionize the materials to be deposited was initiated in the Ginzton Research Center of Varian in 1987, during the same time period as that for the development of the other

I-PVD approaches at Osaka Electro-Communication University,[9] TRW,[10] and IBM.[11] Instead of using a separate plasma source such as RFiPVD or microwave electron cyclotron resonance (ECR) plasma to post-ionize sputtered or evaporated target materials, the HCM relies on the same magnetron plasma to both sputter and ionize the metal atoms. Because only a single DC power supply is required, HCM designs are inherently simpler than other I-PVD approaches.

For a high-power planar magnetron or S-Gun[12] commonly used in semiconductor metalization, the sputtering plasma density is typically $\geqslant 10^{12}\,\text{cm}^{-3}$ with a thickness on the order of the electron Larmor radius (typically about 1 cm)[13] and is located a small distance across the dark space gap from the target surface. In view of the small plasma volume, the high velocity of the sputtered atoms, and the low gas pressure, the relative ionization of sputtered species is typically a few percent or less as they pass through the magnetron plasma.[14]

The use of an intense plasma of a sputter source to ionize the materials to be deposited was demonstrated in the inverted reentrant magnetron (IRMA) ion source.[15] The IRMA source is an inverted post-magnetron ion source designed for ion implantation. By coating or making the cathode with the desired target material, an intense ion beam can be extracted through an aperture in the cathode. Figure 1 shows a typical mass spectrum of the extracted ions from the IRMA source measured by a magnetic sector mass spectrometer. Although the sputtering gas (in this case, argon) makes

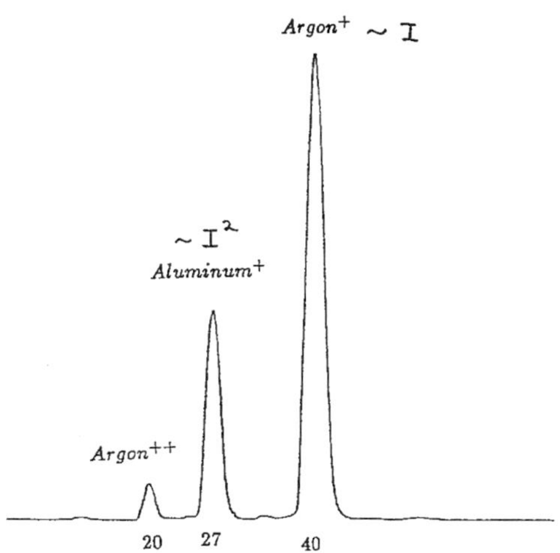

FIG. 1. Extracted ion current versus mass from an IRMA ion source with an aluminum cathode measured by a magnetic sector mass spectrometer (courtesy of Dr. J. C. Helmer).

up the majority of the extracted ions as expected from a typical sputtering discharge, a significant amount of metal (in this case, aluminum) ions can also be generated at high sputtering power. In contrast to the argon ion intensity which has a linear dependence with the sputtering discharge current, the aluminum ion intensity has a square dependence. Thus, it is conceivable that a sufficiently intense sputtering plasma can be used to cause ionization of not only the gas atoms but also many atoms of the material to be deposited.

The IRMA ion source cannot be used as a high-density plasma source because of the lack of an extraction mechanism of the metal plasma within the cylindrical cathode. Poor target utilization (because of the torus-shaped discharge) and cross-field instability at high ion beam current also preclude it as a suitable candidate for I-PVD application in microelectronics. To address the IRMA source shortcomings, the concept of using stacked annular magnets to create a looping magnetic field within a cup-shaped hollow cathode to sustain a magnetron discharge and the use of a leaky magnetic mirror to extract the plasma was first conceived by J. C. Helmer in 1987.[16] A very efficiently cooled cathode was designed and fabricated by G. R. Lavering,[17] using turbulence in the cooling fluid to enhance the cooling efficiency. Although the first HCM was assembled and tested by R. L. Anderson shortly after its creation, its feasibility as a high-density plasma source was not proven until early 1990 by the author.

The newly invented plasma source was referred internally at Varian as a HCM to reflect the confinement of a high-density magnetron plasma within a hollow cathode. It is a simple and accurate description of the essence of the new device. However, this name (HCM) is too generic and confusing due to the existence of many other plasma sources bearing similar or even identical names (e.g., cylindrical hollow cathode/magnetron,[18] cylindrical-hollow magnetron,[19] cold hollow cathode in a gas magnetron,[20] hollow cathode magnetron,[21,22] hollow-cathode magnetron,[23,24] hollow cathode type magnetron,[25] magnetron hollow-cathode discharge,[26] and hollow cathode enhanced magnetron[27]). An alternative and more concise description of the source is probably the "null-field hollow cathode magnetron" as used in the patent application of Ref. 28. To distinguish this high-density plasma source from other devices bearing similar names, we refer to it only as the HCM in the remainder of this chapter.

B. Basic Source Construction

Figure 2 is a schematic cross section of the HCM which illustrates its operating principles and basic construction. By providing a magnetic field

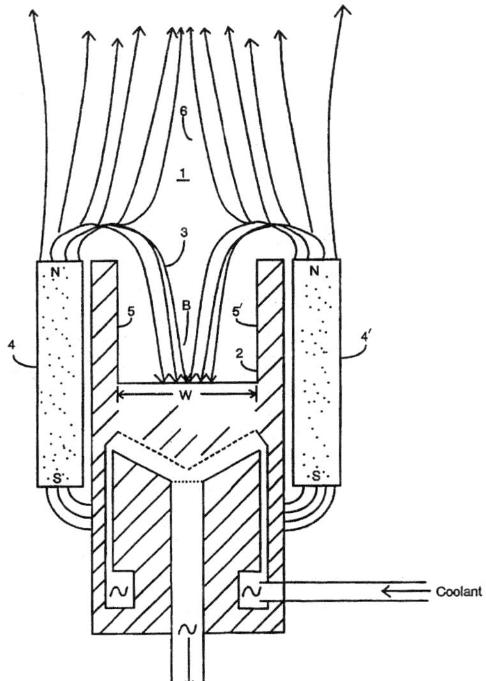

FIG. 2. Schematic cross section of a HCM source [28].

having a magnetic null region (Fig. 2, No. **1**) at the opening of a container (**2**), a magnetic cusp mirror[29] is created that traps electrons inside the container. Other than those electrons which have entered into the magnetic null region at the upper edge (**6**) of the region (**3**) with axial velocity much greater than the radial velocity, additional electrons are reflected back and retained in the container. A stack of permanent magnets (**4** and **4′**) provides magnetic field lines inside of container (**2**) which loop around so that the magnetic field lines are parallel to the container wall surfaces (**5** and **5′**) as in standard sputter magnetron fashion. In a typical HCM, the width and depth of the container are approximately the same. This configuration provides a high probability that sputtered neutral target atoms from surface (**5** or **5′**) are either ionized by the highly intense plasma or redeposit on the opposite wall.

An important distinction that separates the HCM from other hollow cathode-type magnetrons is the use of permanent magnets to create the null field region in front of the cathode opening. Without the use of a plurality of electromagnets, a single electromagnet cannot provide the null field region required for the plasma confinement. A single toroidal magnet (**4** and **4′**) can be used for the cup-like plasma container (**2**), with bar-type magnets (**4** and **4′**) preferred for the configuration in which the plasma container (**2**) is a groove.

By varying the magnetic strength of the main magnet (**4** and **4′**) along the symmetry axis using rotating magnets on the closed end of the hollow cathode, full face erosion of the target surface can be achieved. The plasma and deposition uniformity of the HCM is controlled by the magnetic field strength and the precise shape of the magnetic cusp mirror. Additional permanent magnets or electromagnets beyond the HCM opening can also be used to fine-tune the deposition uniformity of the HCM similar to the magnetic bias[12] scheme used by the S-Gun and the ConMag magnetrons.

As with standard magnetrons, to prevent overheating and resulting deformation of the target, a very efficient cooling system is required to maintain the target at an acceptable equilibrium temperature. A separate, separable or integrated cooling jacket is constructed around the hollow cathode target to remove the heat generated by sputtering. At very high-power operation, the velocity of the water or other suitable coolant flowing in the heat transfer region within the HCM source should be high enough to induce turbulence to maximize the heat transfer efficiency.

C. MAGNETIC CUSP MIRROR

As shown in Fig. 2, the stacked annular magnets create a magnetic null on the radial axis of the hollow cathode at a small distance from the cathode opening. This magnetic null acts as a "cusp mirror" that reflects most of the escaping electrons back into the hollow cathode cavity. By preventing the primary electrons from prematurely escaping from the hollow cathode, a very high-density plasma can be built up within the hollow cathode.

As discussed previously, except for a small percentage of the escaping electrons, most electrons are reflected back into the hollow cathode cavity. To maintain charge balance in the plasma, positive ions are pulled along with the escaped electrons by ambipolar diffusion. As a result, a beam of ionized target atoms is developed and emits from the center of the opening of the hollow confining container. There are several advantages of using a magnetic mirror to extract the plasma. Once the electrons leave the discharge region, the null mirror isolates the beam electrons from coupling with those of the discharge plasma. Thus, the plasma beam formed in a HCM is much more flexible than a standard plasma and can be manipulated and biased without affecting the discharge characteristics of the HCM.

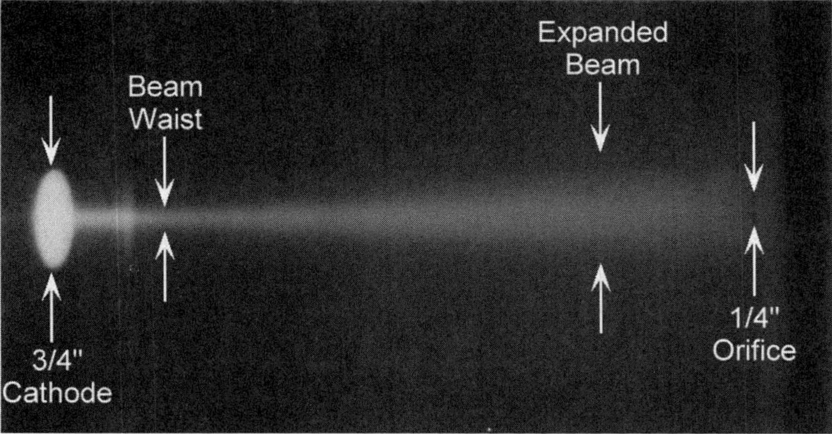

FIG. 3. Digitized image of a plasma beam streaming out of the opening of a $\frac{3}{4}$-in.-diameter HCM toward an electron energy analyzer.

Another valuable property is the quality of the plasma beam. To escape through the loss cone region of a magnetic mirror, the pitch angle of the escaped electrons has to be smaller than a threshold which is in turn determined by the mirror ratio of the cusp mirror.[29] As a result of this phenomenon, the transverse velocity of the plasma beam is greatly reduced. By limiting the beam transverse velocity, the plasma beam becomes much more conducive to steering, focusing, or expanding using externally applied magnetic or electric fields.

Figure 3 shows a digitized image of a plasma beam streaming out of the opening of a $\frac{3}{4}$-in.-diameter HCM. The cusp mirror loss cone forces the plasma to stream only from the central region of the opening along the symmetry axis. The location of the null field region coincides with the "waist" of the beam which is compressed into a much smaller diameter.

D. SOURCE SCALING

To simulate the physics and the ionization mechanism behind the HCM, a simple one-dimensional analytical model was developed and then used to understand the effect of cathode size, plasma density, and electron temperature on the ionization of the sputtered atoms within the hollow cathode. Although this model is simple, it is adequately accurate to provide insight about the scaling and design of the HCM. Due to the high plasma density in the HCM and the low ionization energy of the metal atoms, the dominant ionization mechanism is the one-step electron impact ionization of the target atoms.[30] Thermalization of the sputtered atoms, two-step ionization of metastable species, Penning ionization, and charge exchange are omitted in this much simplified model. More sophisticated plasma models are obviously required for better agreement with experimental measurements.

In the basic configuration of the HCM, the width and depth of the container are approximately the same. Consider a one-dimensional model of our confining cathode in which the height, h, of the container is larger than its width, w. This configuration provides a high probability that the sputtered neutral target atoms will either be ionized by the highly intense plasma or be captured and redeposit on the opposite wall. Since the emission angle of the sputtered target atoms follows approximately a cosine angular distribution, most of the sputtered atoms will either suffer an ionizing collision with a fast electron or travel at least a distance w before hitting the opposite wall. The average number of ionizing collisions, υ_{ionize}, that a sputtered neutral suffers can be described by the following equation:

$$\upsilon_{\text{ionize}} = n_e v_e \sigma \tau_n, \tag{1}$$

where n_e is the average electron density within the hollow cathode, v_e is the average electron velocity, σ is the total ionization cross section by electron impact, and τ_n is the mean lifetime of the neutral before redeposition.

The energy distribution of the sputtered atoms can be approximated by the Thompson model developed by collisional cascade theory.[31] For this model, a mean velocity V_n is used to represent the average velocity of the sputtered neutrals. Thermalization by collision with the background neutrals has also been ignored in this approximation. With this simplification, the average lifetime of the sputtered atoms can be written as

$$\tau_n = \frac{w}{V_n} = w\sqrt{\frac{M_n}{2E_n}}, \tag{2}$$

where M_n and E_n are the mass and average energy of the sputtered neutrals, respectively. Since the electrons are trapped magnetically by the cusp mirror and electrostatically by the closed end of the hollow cathode, we assume that the bulk of the electrons has enough time to interact and reach thermal equilibrium within the hollow cathode. Thus, the electron energy distribution, $f(E)$, can be represented by a Maxwellian distribution of the form

$$f(E) = f(E, T_e) = \frac{e^{-E/kT_e}}{kT_e}, \tag{3}$$

where E is the energy of the electrons. As suggested by Helmer,[32] at low electron energy ($E < 3E_i$), where E_i is ionization energy of the sputtered atom, the total ionization cross section of a neutral atom can be approximated by

$$\sigma(E) \approx \sigma_{\max}\left[1 - e^{-\left(\frac{E-E_i}{E_i}\right)}\right], \tag{4}$$

where $\sigma_{\max}$ is the maximum ionization cross section at the optimum electron energy. By use of Eqs. (2)–(4), Eq. (1) can be rewritten as

$$v_{\text{ionize}} = wn_e\sqrt{\frac{M_n}{mE_n}}\int_{E_i}^{\infty}\frac{e^{-E/kT_e}}{kT_e}\sigma_{\max}(E)\left[1 - e^{-\left(\frac{E-E_i}{E_i}\right)}\right]\sqrt{E}\,dE. \tag{5}$$

With the substitution of the reduced energies,

$$x_j = \frac{E_j}{kT_e}, \tag{6}$$

and the use of the incomplete gamma function, Eq. (5) can be written as

$$v_{\text{ionize}} = wn_e\sigma_{\max}\sqrt{\frac{M_n}{mx_n}}\left[\Gamma(3/2, x_i) - e\cdot\left(\frac{x_i}{x_i+1}\right)^{3/2}\Gamma(3/2, x_i+1)\right]. \tag{7}$$

The fraction of the sputtered neutrals that is ionized by electron impact, f_{ionize}, is related to the average number of ionizing collisions, v_{ionize}, by

$$f_{\text{ionize}} = 1 - e^{-v_{\text{ionize}}}. \tag{8}$$

By examining Eq. (7), it can be seen that v_{ionize} is proportional to the average electron density and the width of the hollow cathode. This result agrees well with the physics of electron impact ionization. Since the electron density of a HCM is approximately proportional to the input power, it is advantageous to operate the HCM at the highest possible power level to increase the ionization fraction. On the other hand, it is possible to scale up the size of a hollow cathode and to reduce the average electron density without lowering the ionization fraction provided that the product of electron density and cathode width remains a constant.

To better understand the effect of electron temperature, we examine the two limiting cases of x_i. In the limit $x_i \to \infty$ (i.e., $kT_e \ll E_i$), Eq. (7) can be approximated as

$$v_{\text{ionize}} \approx wn_e\sigma_{\max}\sqrt{\frac{M_n}{mx_n}}\frac{e^{-x_i}}{\sqrt{x_i}}. \tag{9}$$

At this limit, the fraction of ionized atoms is dominated by the exponential term $-x_i$. A small change in the electron temperature or in the ionization energy has a major impact on the ionization efficiency. The presence of non-Maxwellian tail distribution has a major effect and must be taken account to accurately estimate the ionization fraction.

In the other limit, where $x_i \to 0$ (i.e., $kT_e \gg E_i$), the limiting form of Eq. (7) becomes:

$$v_{\text{ionize}} = wn_e\sigma_{\max}\sqrt{\frac{M_n}{mx_n}}\frac{\sqrt{\pi}}{2}. \tag{10}$$

As can be seen from Eq. (10), once the electron temperature is much higher than the ionization energy, the ionization fraction becomes independent of the electron temperature and the ionization energy. This result is expected since the cross section of total ionization energy of the target atom as in Eq. (4) approaches a constant in this model.

As an example to illustrate the application of the previous ionization model to an actual HCM design, we consider the ionization efficiency of an aluminum HCM used in the earlier experiments. The HCM has a cylindrical cavity with length of 1 in. and an inner diameter of $\frac{3}{4}$ in. An electron energy analyzer was used to measure both the electron density and the temperature of the plasma beam. At 3 kW of input power and 2 mTorr of argon pressure, the electron density and temperature near the waist of the beam were determined to be approximately $10^{13}\,\text{cm}^{-3}$ and 8 eV, respectively. By use of the parameters $n_e = 10^{13}\,\text{cm}^{-3}$, $w = 1.9$ cm, $\sigma_{\max} = 5.4 \times 10^{16}\,\text{cm}^{-3}$, $E_i = 5.985$ eV, $E_n = 10$ eV, and $kT_e = 8$ eV, the ionization efficiency, f_{ionize}, is calculated to be 54.8%.

In the previous estimate, more than half of the sputtered aluminum neutrals have been converted to aluminum ions in the HCM. Since the plasma density and electron temperature within the hollow cathode are expected to be much higher than those near the waist of the plasma beam and most of the un-ionized target material is captured inside the hollow cathode, the actual percentage of ionized metal flux arriving at the substrate is expected to be much higher than that calculated from Eq. (8).

III. Source Characterization

A. Operational Characteristics

For a typical planar magnetron of optimum field shape and intensity, the voltage–current curves follows the relationship

$$I = kV^n, \tag{11}$$

where I is the cathode current and V is the cathode potential.[33] The higher the exponent n, the more efficient the electron trapping in the plasma. Typical V–I characteristics for a $\frac{3}{4}$-in.-diameter Al HCM are shown in Fig. 4 for various argon pressures. As can be seen, the V–I characteristics of a HCM follows approximately the relationship as described in Eq. (11) at low sputtering power. Depending on the argon pressure, the exponent n ranges from 7.8 at 2 mTorr to 27.8 at 20 mTorr which is significantly higher than those of a planar magnetron indicating the excellent electron trapping by the magnetic cusp mirror. At sufficiently high sputtering power and particularly low argon pressure, the discharge current of the HCM increases much slower with discharge voltage than in the V–I relationship as described in Eq. (11). This phenomenon can be attributed to the gas density reduction effect by the high-density plasma and the energetic species within the hollow cathode.[34] The higher the discharge power and the lower the gas pressure, the more pronounced is the gas density reduction by the high-density plasma. For a small cathode that has limited gas conductance, it is advantageous to introduce gas directly into the cathode to minimize the gas density reduction for very high-power, low chamber pressure operation. Because of its high plasma density and efficient plasma confinement, the impedance of the HCM is much lower than that of a typical magnetron of its size. At sufficiently highly pressure and high power of operation, the HCM tends to run in a constant voltage mode in which the operational voltage is almost independent of the input power.

The HCM can be used for the deposition of a large variety of conductive materials simply by changing the sputtering target. Pulsed DC or RF can be used for the sputtering of nonconductive materials in a similar manner

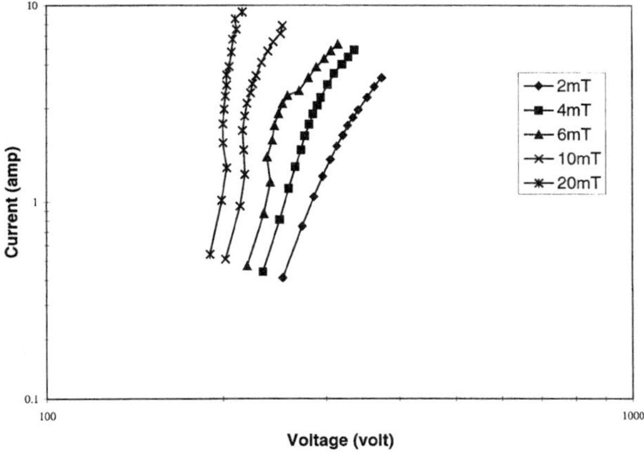

FIG. 4. V–I Characteristics of a $\frac{3}{4}$-in.-diameter Al HCM at various chamber argon pressures.

to that in other sputtering magnetrons.[33] By use of a mixture of inert and reactive gases such as nitrogen or oxygen, many different kinds of binary compounds (e.g., TiN, TaN, Ta_2O_5) can be deposited using reactive sputtering.[35] Because of its extremely high plasma density and high sputtering rate, the HCM is capable of maintaining the target surface in the non-poisoned (i.e., the metallic) mode, avoiding many drawbacks commonly encountered in reactively sputtering.

B. Plasma Parameters

1. *Electrical Diagnostics*

Due to the very high plasma density and the difficulty of inserting physical probes within the hollow cathode without adversely disturbing the discharge, we measured the plasma density and electron temperature of the HCM only beyond the magnetic null region using a movable electron energy analyzer and Langmuir probes. Since the plasma parameters measured by the Langmuir probes are similar to those measured by the electron energy analyzer and have been presented elsewhere,[36] only the results of the electron energy analyzer are discussed in details in this section.

A movable electron energy analyzer was used in place of the substrate holding stage to measure the plasma density and electron temperature of the HCM plasma as a function of distance along the symmetry axis. An orifice plate of variable orifice diameter was used instead of a metal grid (commonly used in other analyzers) as the front electrode of the analyzer to minimize the reduction of collection efficiency by metal deposition. To avoid overheating by the high-density plasma, the orifice plate is made of copper and attached to a water-cooled housing which is movable and electrically isolated to simulate the holding stage and to minimize the perturbation to the plasma. A second copper plate placed at a small distance behind the orifice serves as the collecting electrode of the analyzer. By varying the collector voltage (with respect to the chamber ground), the electron energy distribution function and thus the electron temperature can be measured with minimal disturbance to the measured plasma. The potential on the first electrode was monitored to verify that the plasma floating potential was not affected by the collector bias. By applying a negative voltage on the collector (typically -50 V), all the ions entering the orifice were collected. The saturated ion current is independent of the distance between the orifice plate and the collector, indicating that the measured ion current density (through the orifice) represents the ion density similar to that collected on a floating conductive substrate. In essence, the orifice plate can be treated as an ideal planar Langmuir probe.

The electron energy analyzer measurement was conducted on a $\frac{3}{4}$-in.-diameter Al HCM. Figure 5 plots the collected ion current of the analyzer as a function of magnetron power for three different orifice diameters (D_c). At low magnetron power, the collected ion current and thus the plasma density increase linearly with the magnetron power. Depending on the size of the orifice, the collected ion current tends to saturate at sufficiently high plasma density due to the breakdown of the sheath as the sheath thickness becomes smaller than the orifice diameter.[37] In order to obtain ion counts that are proportional to the ion density, the size of the orifice must be made small but remain large enough to avoid clogging by the deposited metal.

Figure 6 presents the collected electron current as a function of the collector voltage at a fixed distance of 2.4 in. from the HCM at three different chamber argon pressures. To ensure a linear dependence of collected ion current with plasma density, the diameter of the orifice and the magnetron power were kept at $\frac{1}{4}$ in. and 0.5 kW, respectively. The electron temperatures were deduced from the slopes of the collected electron current and determined to be 7.7, 6.3, and 4.9 eV for 2-, 8-, and 20-mTorr argon pressures, respectively. The radial profile of the emitted plasma beam can be inferred by varying the distance (L_c) between the electron energy analyzer and the hollow cathode. Figure 7 shows the collected ion current as a function of distance from the hollow cathode for three sizes of orifice. At a distance close enough to the cathode opening ($L_c < 1.5$ in.), the collected ion current tends to be a constant as the size of the orifice becomes larger than the plasma beam. If the plasma spreads linearly with distance, the ion current will decay following the inverse square law. This hypothesis has been

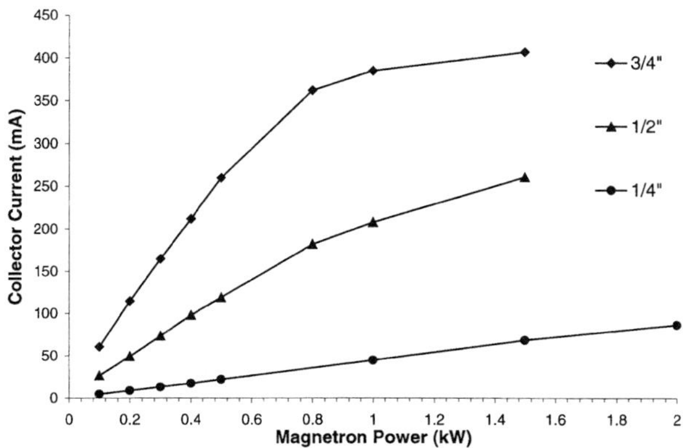

FIG. 5. Collected ion current of the electron energy analyzer as a function of magnetron power for three different aperture sizes ($P_o = 2$ mTorr, $V_c = -50$ V, $L_c = 2.4$ in.).

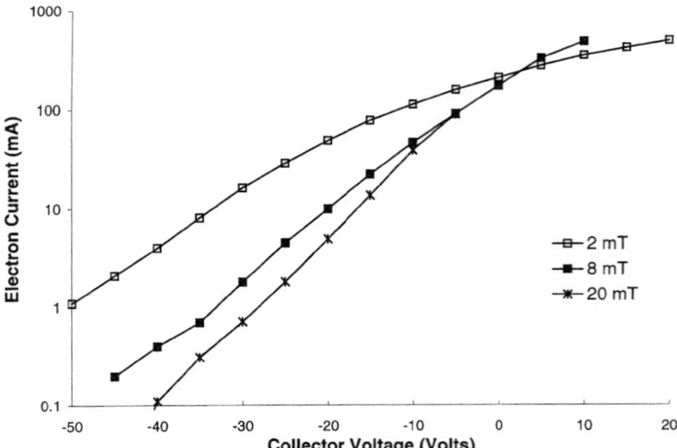

FIG. 6. Collected electron current of the electron energy analyzer as a function of the collector voltage at three different chamber argon pressures ($P = 0.5\,\text{kW}$, $D_c = \frac{1}{4}$ in., $L_c = 2.4$ in.).

verified as the inverse square root of the ion current density increases linearly with the increase of the distance for $L_c \geqslant 3.0$ in. By extrapolating the linear portion of the curves toward the x-axis, a common x-intercept can be found at $L_c \sim 0.28$ in. which coincides with the location of the magnetic null. Again, this illustrates the effect of the magnetic null in the extraction of the plasma and its role as a single aperture lens similar to a sampling orifice that determines the divergence of the extracted plasma beam.

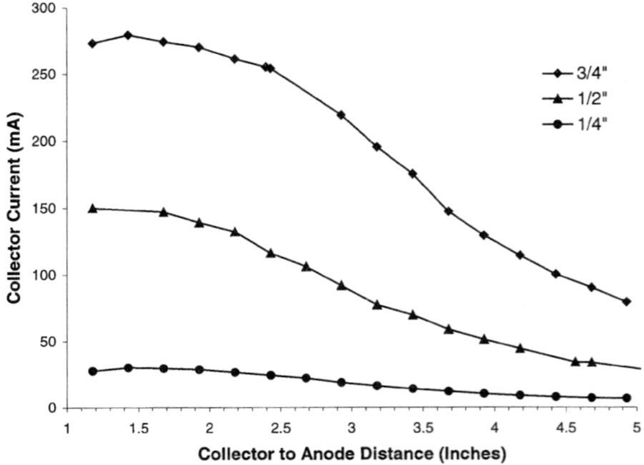

FIG. 7. Collected ion current of the electron energy analyzer as a function of collector to anode distance L_c, for three different aperture sizes ($P_o = 2\,\text{mTorr}$, $V_c = -50\,\text{V}$, $P = 0.5\,\text{kW}$).

2. *Optical Diagnostics*

Optical emission spectroscopy (OES) and optical emission imaging (OEI) have been used to characterize the HCM plasma. Since these two diagnostics are nonintrusive, it is possible to measure the properties of the HCM plasma both within and outside of the hollow cathode discharge region. Due to the complexity in the physics and data analysis of the plasma emission, however, only qualitative results can be deduced easily from these measurements.

OES is a very useful technique for understanding the ionization and excitation processes in the HCM plasma. By measuring the emission lines of the various species of interest, the relative ratio of ionization of the metal and the inert gas atoms and the relative density among the various species can be deduced using actinometry or other more sophisticated ionization models. Since OES is not an absolute measurement technique, a second diagnostic is needed to calibrate the OES before it can provide any quantitative information.

Figure 8 presents the normalized emitted light intensity for all the major emission lines from 390 to 440 nm as a function of the magnetron power for

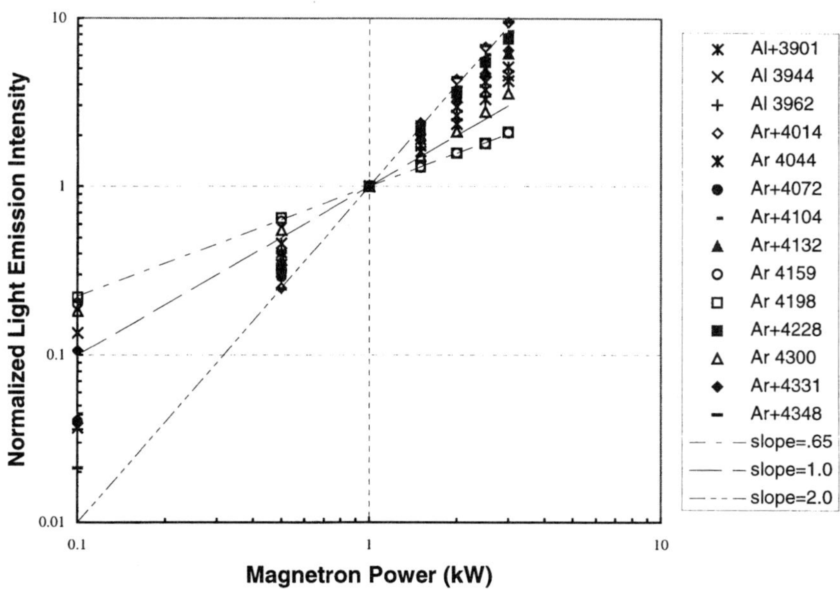

FIG. 8. Normalized emitted light intensity for all the major emission lines from 390 to 440 nm as a function of the magnetron power for a $\frac{3}{4}$-in.-diameter Al HCM.

a $\frac{3}{4}$-in.-diameter Al HCM. The intensity of nearly all the emission lines follows a power law dependence with the magnetron power. This phenomenon can be explained by a simple cascade theory on the ionization and excitation of the various species. As shown in the previous section, the electron density, n_e, of the HCM plasma increases linearly with the magnetron power. Since the number of argon neutral atoms is approximately constant in the discharge, the rate of excitation of the argon atoms by electron impact should be proportional to the electron density. Thus, the exponent of the argon neutral emission lines, Ar^* with the magnetron power should be close to unity. On the other hand, the ionization and excitation of the argon ion emission lines, $(Ar^+)^*$ by electron impact can be due to either the single-step ionization and excitation as described by the equation,

$$Ar + e^- \rightarrow (Ar^+)^* + 2e^-, \tag{12}$$

or by the two-step ionization and excitation as described by the equations

$$Ar + e^- \rightarrow Ar^+ + 2e^-, \tag{13}$$

$$Ar^+ + e^- \rightarrow (Ar^+)^* + e^-. \tag{14}$$

The branching ratio of these two excitation mechanisms is controlled by the electron energy distribution. Depending on the electron temperature and the excitation energies of the specific argon emission lines, the exponent is between 1 and 2. Since the argon ionization energy is much higher than the electron temperature under normal deposition conditions, the dominant mechanism for the generation of the excited argon ions is expected to be the two-step process, which gives an exponent of slightly less than 2.

Optical emission by metal neutrals and ions can be described in a three- or four-step process. First, the argon ions are created by electron impact ionization as in Eq. (13). Next, the metal atoms are sputtered off the target by the argon ions which can be written as

$$Ar^+ + M_{(s)} + e^- \rightarrow Ar + M_{(g)}. \tag{15}$$

Finally, the metal atoms are excited by electron impact given by the equation

$$M_{(g)} + e^- \rightarrow M^* + e^-. \tag{16}$$

Similar to the argon ion emission, excited metal ions can be generated by either a single- or two-step electron impact ionization and excitation process. The single-step process can be represented as

$$M_{(g)} + e^- \rightarrow (M^+)^* + 2e^-, \tag{17}$$

whereas the two-step process can be represented by the equations

$$M_{(g)} + e^- \rightarrow M^+ + 2e^- \quad (18)$$

$$M^+ + e^- \rightarrow (M^+)^* + e^-. \quad (19)$$

Due to the low ionization and excitation energies of most metal atoms, the dominant mechanism for the generation of excited metal ions is the single-step process. In the case in which only a small percentage of the metal atoms are ionized, the exponents for the excited metal neutrals and ions are 2 and 3, respectively. In contrast, if most of the metal atoms are ionized and excited by the high-density plasma immediately, the exponents are reduced to 1.

As can be seen in Fig. 8, the measured exponents for Ar^* and $(Ar^+)^*$ respectively ranged from 0.65 to 1.4 and from 1.6 to 2.0, whereas those for Al^* and $(Al^+)^*$ were 1.5 and 1.6, respectively. Other than a few discrepancies between the measured and the predicted values, the general trend of the measured exponents agrees well with the cascade ionization and excitation theory. The measured values of 1.5 and 1.6 for the Al^*, and $(Al^+)^*$ exponents indicate that substantial amounts of the Al atoms have been ionized within the hollow cathode. The lower than expected exponents for most of the Ar^* and $(Ar^+)^*$ lines are probably due to the heating and rarefaction of the argon neutrals by the high-density plasma. On the other hand, the exponents of some of the Ar^* lines are found to be significantly higher than unity. A plausible explanation is the effect of the argon metastable states. Depending on the excited state of a specific emission line, some of the emission lines may also be excited from an argon metastable state via a two-step process.

OEI is a complimentary technique to OES used to obtain the spatial distribution of a particular optical emission line. OEI is performed by placing a narrow bandpass interference filter in front of a CCD camera. By carefully selecting the wavelength of the bandpass filter to match the emission lines of interest, the spatial distribution of a particular species can be inferred from the measured light emission profiles. Figures 9a and 9b show the contour map of the light emission intensity at 420 nm of the HCM plasma beam (side view) without and with the presence of a floating substrate, respectively. The waist of the beam is at the left-most side whereas the substrate holder (if present) is at the right-most side. In the absence of the substrate as shown in Fig. 9a, the plasma beam spreads linearly with distance from the cathode. Figure 10 shows the radial profiles of the emitted light at various positions along the symmetry axis. As can be seen, the plasma emission profiles approximately resemble a Gaussian and spread linearly as the plasma beam streams away from the cathode opening. In the

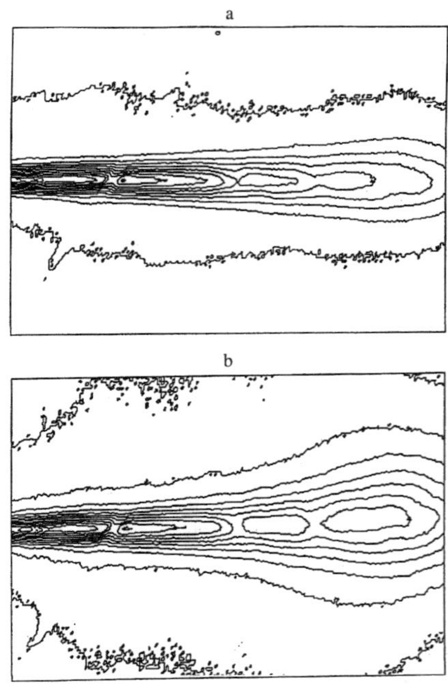

FIG. 9. Contour map of the light emission intensity at 420 nm of the HCM plasma beam (side view) (a) without a substrate and (b) with a substrate.

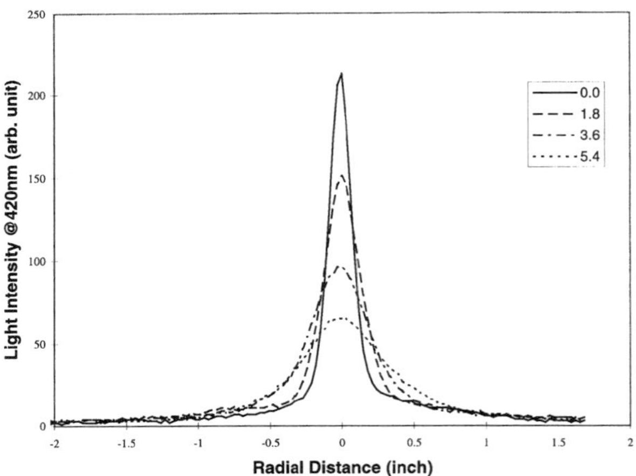

FIG. 10. Radial profile of the plasma emission at 420 nm at various distances from the HCM opening.

presence of a floating substrate, the plasma beam tends to hover around the surface of the substrate as shown in Fig. 9b due to the repulsion of electrons by the negatively charged substrate. Incomplete shielding of the substrate electric field in the presheath is evident by the much increased beam divergence at a distance far from the substrate.

C. UNIFORMITY AND DIRECTIONALITY CONTROL

Beyond the magnetic null region, the remnant-reversing magnetic field beyond the magnetic null region serves to guide and collimate the extracted plasma beam from the magnetic null to the substrate. The strength of this remnant-reversing magnetic field determines the rate of spreading of the plasma beam and the uniformity of the ion density across the surface of the substrate. By engineering and controlling the strength and shape of the remnant-reversing field, it is possible to achieve excellent plasma, ionization, and deposition uniformity across a large substrate using a relatively small-size hollow cathode. For example, an 82-mm-diameter HCM can provide film thickness uniformity of better than 3% (1σ) on a 125-mm-diameter wafer. In contrast to a planar magnetron in which the deposition uniformity is determined by the target erosion profile and the source to substrate spacing (SSS), the uniformity of HCM deposition is not sensitive to the target shape, the erosion profile, or the SSS primarily due to the very high ionization efficiency of the metal atoms within the hollow cathode.

As discussed for the much simplified ionization model, the degree of ionization of the metal atoms in a HCM is controlled by the cathode size, plasma density, electron temperature, ionization energy, and electron impact cross section of the metal species. In practice, the inert gas pressure, magnetron power, and SSS are the basic knobs that can be varied during the deposition process. In addition to the degree of ionization, the directionality of the deposited flux on a substrate is also controlled by the effective bias on the substrate and the transverse energy of the incoming ions. In general, the higher the substrate bias, the more directional is the incoming ions. Excessive substrate bias beyond a threshold voltage (typically -50 V), however, is problematic due to inert gas incorporation, ion-induced defects, and excessive resputtering of top corners. Charging damage to the underlying gate dielectric is also a concern for the metallization of the first level contacts. Effective substrate bias can be achieved either by electrically isolating the substrate to let it self-bias to the floating potential of the incoming plasma or by biasing it with a separate power supply on the holding stage. Since most semiconductor devices have nonconductive coatings, including oxide or other dielectrics, RF or pulsed

DC bias are generally needed for consistent external bias control. The major concern of applying an external bias to the substrate is the possibility of charging damage and the potential for arcing between the substrate and the holding stage. With a proprietary design, we have found that the HCM can generate an adjustable self-bias of -10 to -60 V on a floating substrate, which has proven to be adequate to impart a high degree of directionality for most deposition applications.

IV. Process Results

In this section, some deposition results for semiconductor metallization applications using the HCM source are presented. All the depositions were performed on a floating substrate without applying a separate bias to the holding stage. In addition to the change of the sputtering target and minor adjustment of the process parameters, the deposition of all the different materials reported here can be accomplished using the same HCM source. Due to the rapid development of the HCM technology and the proprietary nature of specific process recipes and capabilities, these results are intended to show the feasibility of using the HCM for various applications rather than the state of the art. Therefore, some of the reported results may be outdated and superseded by new processes developed after this book was published.

A common way to characterize the directionality of a deposition process in VLSI metallization is to measure the percentage of bottom (B) and sidewall coverages (S) with respect to the field thickness of the deposited film on patterned features of a given aspect ratio (AR, defined as the feature depth divided by the feature width or diameter). The features can be of the shape of a trench, a circular hole (commonly referred to as contacts or vias depending on the underlying layer), or a via within a trench (commonly referred to as a dual-damascene structure). Assuming a conformal coating (i.e., $B = S$), a unity sticking coefficient, and the conservation of mass, it can be easily shown that the step coverages for trenches and vias for conventional PVD (including collimation and long-throw depositions) are related to the AR by the equations

$$S_{\text{trench}} = \frac{1}{1 + 2 \cdot AR}, \tag{20}$$

$$S_{\text{via}} = \frac{1}{1 + 4 \cdot AR}. \tag{21}$$

Since B is larger than S for most deposition processes, Eqs. (20) and (21) are in practice the upper limits for the minimum sidewall coverages which decrease inversely with the increase of *AR*. Since a trench has a narrow width only in one dimension while a via is a two-dimensional structure, the bottom and sidewall coverages of a trench can be twice as much as those of a via. Similarly, a dual-damascene structure has step coverage somewhere in between a trench and a via (e.g., the effective *AR* of a 3:1 via under a 3:1 trench is higher than a 3:1 but much less than a 6:1 via).

One major source of confusion in comparing the directionality of different deposition processes (particularly if they are offered by different vendors) is the lack of a standard for the feature width used in computing the *AR*. Depending on the tapering of the sidewall and the exact profile of the feature, it is not uncommon to have a factor of two variation in the opening and the contact widths and thus the computed *AR* for a given feature. The measured opening and contact widths can vary substantially even for the same type of feature due to beveling at the top and rounding near the bottom. Thus, a better and more robust definition which we used is to compute the *AR* by dividing the feature depth with the feature width at the half-depth.

A. Ti/TiN LINERS

PVD titanium (Ti) and titanium nitride (TiN) are widely used as contact liners to improve contact/via resistance, as adhesion/diffusion barriers for chemical vapor deposition (CVD) tungsten, and also as wetting layers for PVD aluminum plugs. With the continuing shrinkage of feature sizes and increasing contact/via aspect ratios, it is becoming more difficult for conventional or even collimation to provide adequate bottom and sidewall coverages. Although CVD Ti and TiN have been shown to provide excellent conformality in high aspect ratio features, the presence of halide and carbon impurities, high resistivity,[38] and a parasitic resistance penalty[39] (due to excessive sidewall coverage) have generated concerns over their uses in next-generation devices. CVD TiN, in particular, has a resistivity ($>100\ \mu\Omega$-cm) significantly higher than that of collimated PVD TiN ($\sim 45\ \mu\Omega$-cm) currently in production. With the further reduction in feature size and the move from tungsten to aluminum plugs, it is highly desirable to lower both the resistivity and the thickness of the Ti/TiN liner.

1. Ti/TiN Film Characterization

The HCM source operates in two distinct TiN deposition modes similar to what has been observed for collimation,[40] namely, the nitrided mode (NM), where the target surface is fully nitrided, and the nonnitrided mode (NNM), where the surface is not nitrided. Details of the experiment has been described elsewhere.[41] As seen in Fig. 11, the resistivity of the HCM TiN film increases monotonically with increasing nitrogen flow. Unlike collimated deposition, which has a sharp transition and hysteresis between the NM and NNM modes, the HCM source exhibits a smooth and continuous shift in process parameters. Only a small percentage (10–20%) of nitrogen is required for NNM TiN deposition (vs. ~50% for collimated TiN). The deposition rate of the NNM TiN is ~20% higher than that for Ti, and a factor of three more than that for the NM. The density of the NNM TiN film was determined either by using Rutherford backscattering (RBS) (to derive the total number of Ti atoms per unit area) and X-ray reflection (XRR), or X-ray fluorescence (XRF) and profilometry. XRR also gives a density number directly, which is 2–4% higher than the 5.22 g/cm^3 canonical bulk value. Combining RBS with XRR yields a density 8–10% above bulk. The Ti/N ratio is found to be 1.03 ± 0.06 by RBS. Within experimental errors, the HCM NNM TiN is determined to be stoichiometric with a density equivalent to that of the bulk material.

Figure 12 shows the NNM TiN resistivity decreasing with increasing deposition temperature. This dependence becomes less important as the substrate temperature is increased above 200°C. In a separate test using a high-purity Ti target, the resistivity of the TiN films on SiO_2 is calculated to be as low as 21 $\mu\Omega$-cm (measured at ambient temperature). This is significantly lower than that of other polycrystalline TiN films and approaches the value (18 $\mu\Omega$-cm) of single-crystal TiN.[42] Unlike other reactively sputtered TiN whose resistivity is observed to increase by as much as 8% when

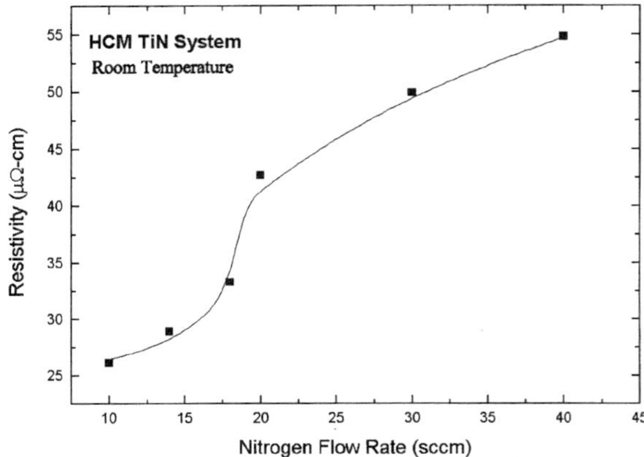

FIG. 11. Correlation of HCM TiN film resistivity with nitrogen flow rate [41].

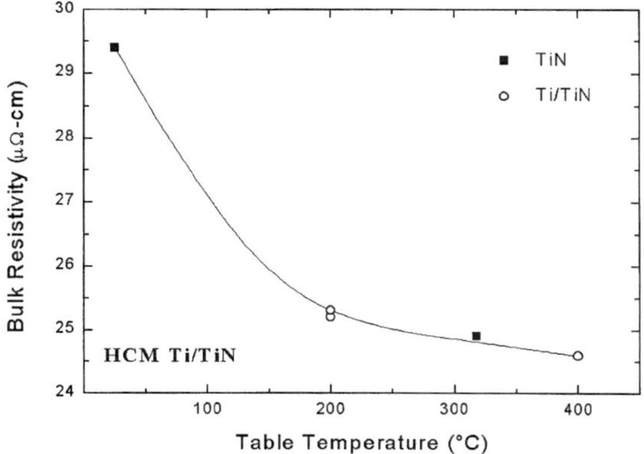

FIG. 12. Correlation of HCM TiN film resistivity with deposition temperature [41].

exposed to room ambient, these TiN films, with thickness ranging from 300 to 3000 Å, have exhibited no measurable change in resistivity over a period of several months.

Figure 13 compares the resistivity of Ti and TiN deposited by the HCM and collimated sputtering as a function of film thickness. As shown in the figure, the resistivities of both the HCM and collimated Ti are nearly identical, whereas the resistivity of the NNM HCM TiN deposited at 200°C is more than a factor of four lower than that for the NM collimated TiN deposited at room temperature. As shown in Fig. 14, the stress of the HCM NNM TiN films as deposited at room temperature is highly compressive,

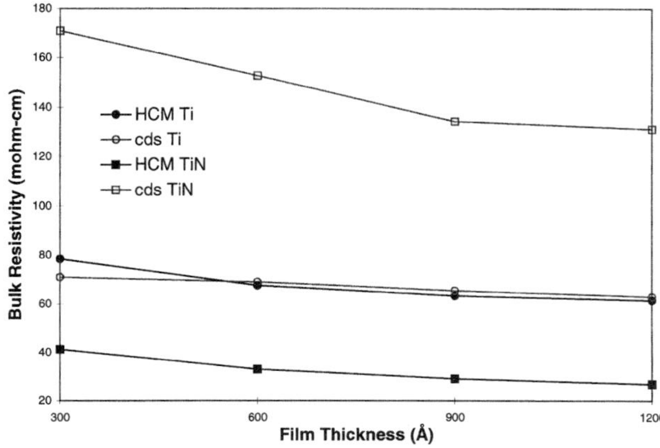

FIG. 13. Thickness dependence of the resistivity of the HCM and collimated Ti and TiN [41].

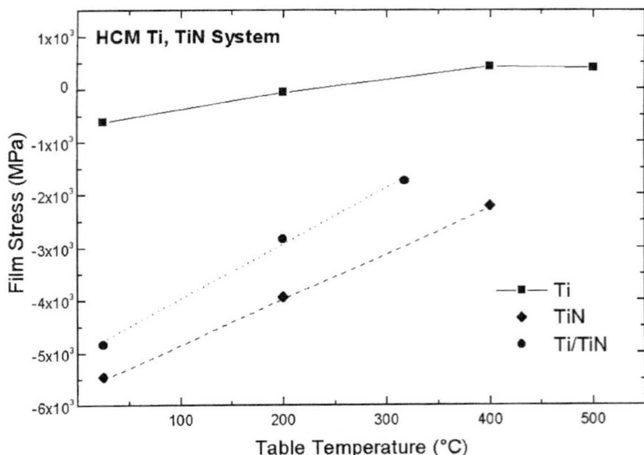

FIG. 14. Correlation of HCM Ti, TiN, and Ti/TiN film stress with deposition temperature [41].

but can be lowered to comparable levels to collimated TiN by using a Ti underlayer and higher deposition temperature ($>200°C$).

The preferred orientation and grain size of the Ti and TiN films are determined by X-ray diffraction (XRD). As can be seen from the θ-2θ coupled scans (Figs. 15–17), the HCM Ti and Ti/TiN films are highly oriented with nearly 100% Ti(002) and TiN(111) peaks, respectively. In the absence of a Ti underlayer, a weak TiN(200) peak is observed. The dramatic change in the preferred orientation of the HCM TiN films can be explained by the close match between Ti(002) and TiN(111) lattice constants. The average grain sizes of the Ti and TiN films are estimated from the full width at half-maximum (FWHM) of the coupled scan according to Scherrer's equation modified with Jone's correction for the instrument broadening.[43]

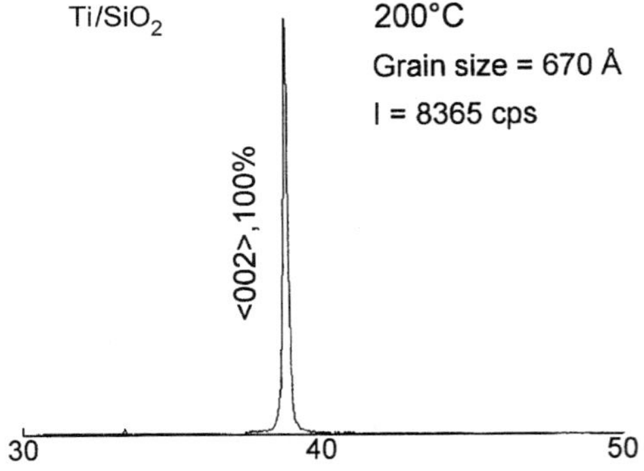

FIG. 15. XRD θ-2θ coupled scan of HCM Ti film on SiO_2, $t = 950$ Å [41].

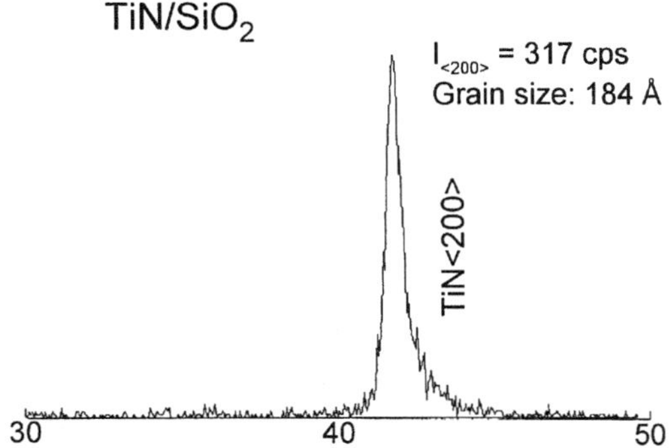

FIG. 16. XRD θ-2θ coupled scan of HCM TiN film on SiO_2, $t = 2280$ Å [41].

Since TiN exhibits columnar grain growth, the estimated grain size is in the perpendicular direction, which can be extremely large compared to the column diameter. These estimates, however, are lower limits of the grain size since lattice parameter fluctuations due to strain or composition variations add to the line broadening. In consideration of these effects, the grain size is probably comparable to the total film thickness and the grains are crystallographically coherent.

The grain size in the parallel direction (column diameter) is determined by TEM. Figures 18 and 19 show the plan view together with the electron diffraction pattern of the HCM TiN with and without a Ti underlayer. Again, the column diameter of the Ti/TiN film is significantly larger than that of the TiN alone. The diffraction patterns also confirm the dramatic differences in the preferred orientation between these two films. As can be seen from Figs. 20a and 20b, the HCM TiN film is very smooth ($\sigma = 4.28$ Å)

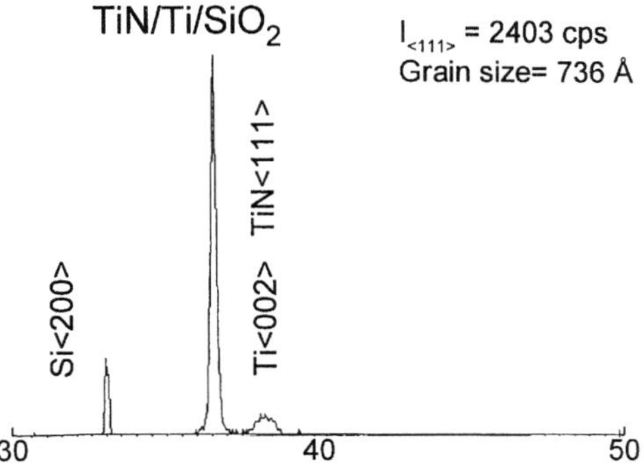

FIG. 17. XRD θ-2θ coupled scan of HCM Ti/TiN film on SiO_2, $t = 1200$ Å [41].

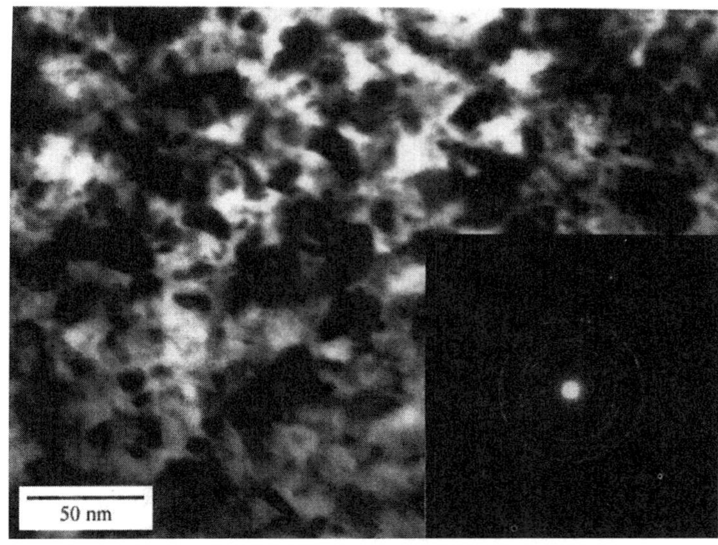

FIG. 18. TEM plan view of HCM TiN film with diffraction pattern, 0° tilt [41].

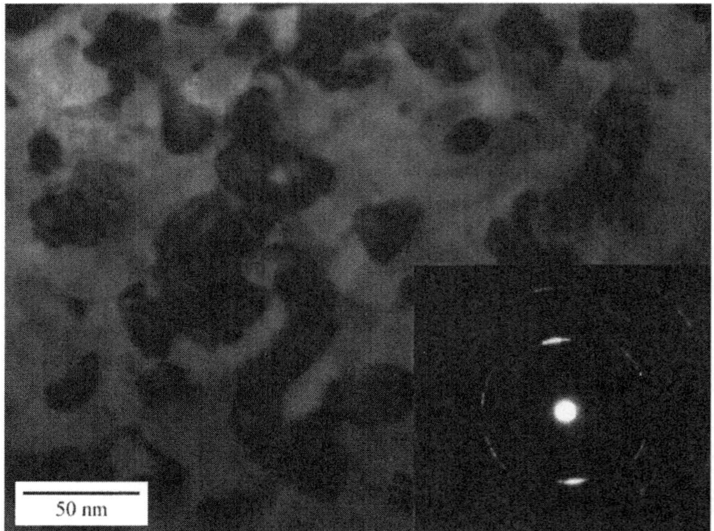

FIG. 19. TEM plan view of HCM Ti/TiN film with diffraction pattern, 0° tilt [41].

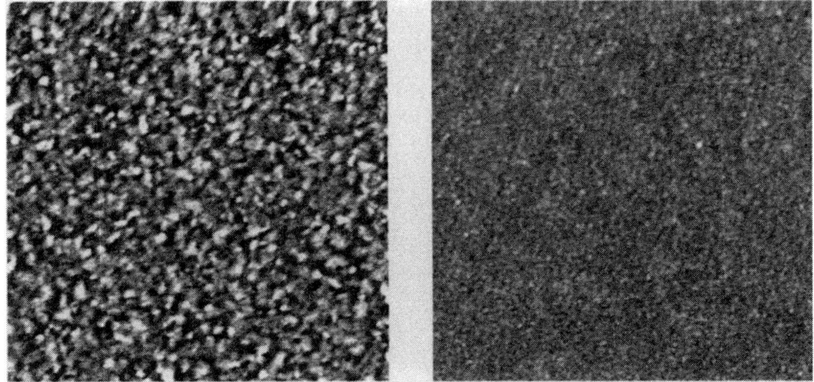

FIG. 20. AFM plan views of (a) collimated Ti/TiN and (b) HCM Ti/TiN films, on SiO_2, image cover $(2\,\mu m)^2$ [41].

and has very tight grain boundaries, in comparison to a standard collimated Ti/TiN film ($\sigma = 11.63$ Å). The gray scales in the two AFM pictures are set so that a given shade corresponds to the same relief in both images. XRR and reflection high-energy electron diffraction also corroborate the AFM results regarding the film smoothness.

Figures 21a and 21b present the cross-sectional TEM views of HCM Ti/TiN film deposited without applied wafer bias on 0.27 μm-diameter contacts of aspect ratio 4.5. The bottom coverage of HCM Ti/TiN is flat and much greater than that of collimated Ti/TiN. Excellent sidewall (minimum 12%) and bottom (maximum 33%) coverages are obtained. As can be seen in Fig. 21b, near conformal coating is achieved even around the bottom corner where the contact is slightly reentrant due to overetching. The minimum sidewall coverage for HCM Ti/TiN deposition is more than a factor of three higher than that expected for conventional PVD based on conservation of mass arguments. This discrepancy can probably be accounted for by nonunity sticking coefficient, angle-dependent resputtering, and/or increased mobility due to high-flux low-energy ion bombardment.

2. *Integration of HCM Ti/TiN With Low-Pressure Al Planarization*

The low-pressure (LP) Al planarization (or MaxFill) process is a promising technology that has been shown to be effective in filling 0.35-μm structures with aspect ratio $\geqslant 3$.[44] The MaxFill process requires an underlayer that is conformal and has good Al wetting properties. Although collimated Ti works well as an Al wetting layer, the formation of high-resistivity $TiAl_x$ at

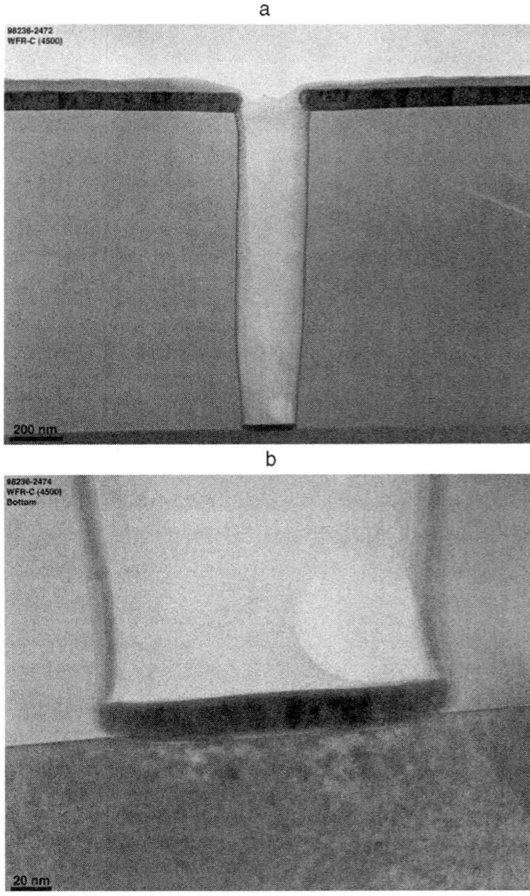

FIG. 21. TEM cross-sectional view of HCM Ti/TiN film on 0.27-μm-diameter, aspect ratio 4.5 contacts (a) overall and (b) bottom corners (courtesy of Dr. C. D'Couto).

the Ti and Al interface is a concern for sub-0.35-μm vias. While far less reactive than Ti, collimated TiN is not a suitable underlayer because of its poor Al wetting properties. Feasibility of treating the TiN surface with ECR plasma to enhance its Al wettability has been reported.[45]

Details of the following experiment have been described elsewhere.[46] By depositing PVD Al-0.5% Cu alloy on top of the HCM Ti/TiN film, a highly oriented Al(111) film can be obtained. From the XRD rocking curve in Fig. 22, the Al(111) FWHM was determined to be 0.54°, which was significantly smaller than the best results deposited on collimated or RFiPVD Ti/TiN

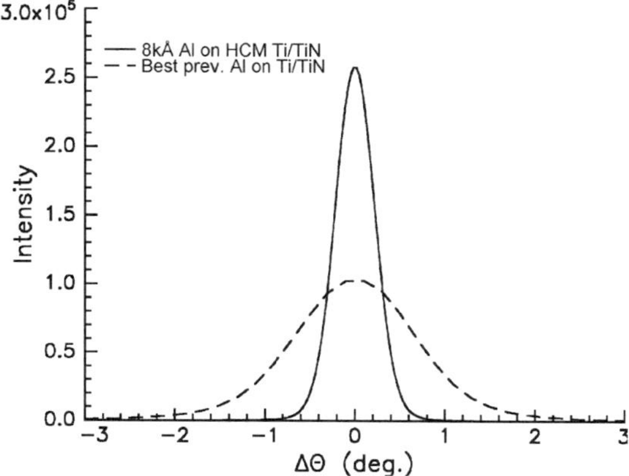

FIG. 22. Comparison of Al(111) rocking curve of PVD Al-0.5% Cu on HCM and collimated Ti/TiN [41].

film.[47] Since a strong Al(111) preferred orientation is most resistant to electromigration failure,[48] the HCM Ti/TiN bilayer is a promising liner for Al metalization.

Figure 23 shows the importance of the underlayer to the grain orientation of the LP Al. As can be seen, the XRD intensity of the Al(111) peak varies 2.5 orders of magnitude between the Al deposited on HCM Ti/TiN and SiO_2. It should be noted that the sequence of the deposition was of paramount importance to the grain orientation as evident by a factor of 100

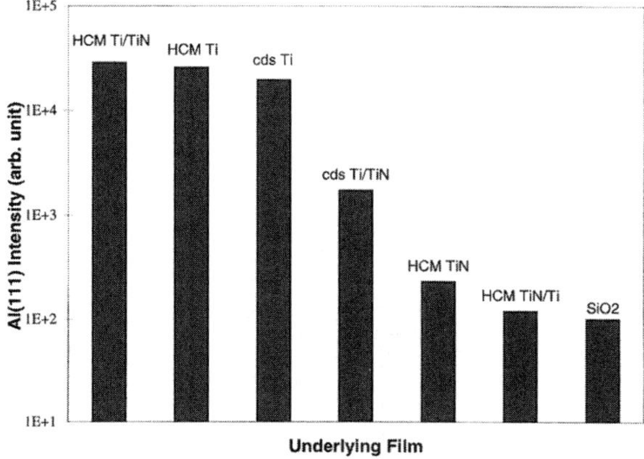

FIG. 23. XRD intensity of Al(111) peak of 7 kÅ of Al-0.5% Cu on different types of underlayers [46], © 1998 IEEE.

difference in the Al(111) intensity between HCM Ti/TiN and HCM TiN/Ti underlayers. We attributed this to the close lattice constant match of Ti(002) and TiN(111) which allowed the Ti to serve as a template for subsequent Al(111) growth.

The effect of Ti thickness to the Al film reflectivity in a Ti/TiN film stack is depicted in Fig. 24. The thicknesses of the TiN and Al are 600 Å and 7 kÅ, respectively, while the Ti thickness varies from 50 to 600 Å. Since HCM TiN is a good Al wetting layer, the LP Al reflectivity is only weakly dependent on the Ti underlayer thickness. In contrast, the LP Al reflectivity increases steadily with the increase in the collimated Ti thickness illustrating the importance of the Ti template effect.

Figure 25 illustrates the dependence of the LP Al reflectivity with the film surface roughness. In general, the smoother the Al surface, the higher the reflectivity. The surface roughness of MaxFill Al using HCM Ti/TiN is similar to that of Al films deposited on a collimated Ti underlayer.

The stress of the Ti/TiN/Al stack as a function of the Ti underlayer thickness for 7 kÅ of LP Al on 600 Å of TiN is shown in Fig. 26. The stress of the Al film on SiO_2 is tensile (~ -200 MPa). With the HCM Ti/TiN underlayer, it becomes slightly compressive. Since the stress of 600 Å of HCM TiN is ~ -6000 MPa, the stress of the stack (~ -80 MPa) is significantly less than that of the individual films combined. This effect is even more pronounced for the collimated Ti/TiN which is compressive (~ -3500 MPa) while the stress of the stack becomes more tensile. Again this illustrates the effect of underlayer on the LP Al structure and the possible annealing of the Ti/TiN underlayer during Al deposition.

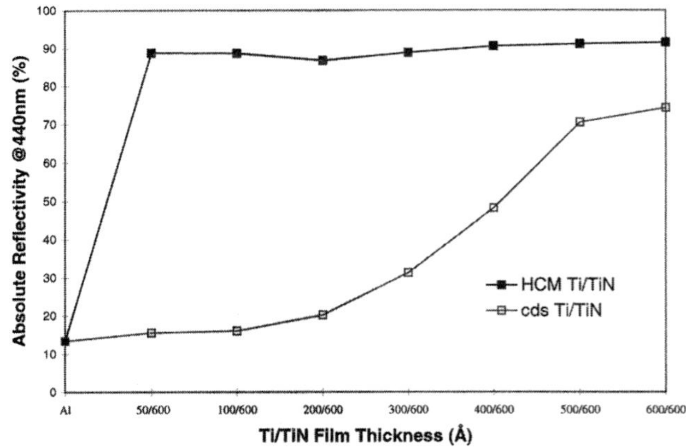

FIG. 24. Reflectivity of LP Al on HCM and collimated Ti/TiN underlayers of various Ti thickness [46], © 1998 IEEE.

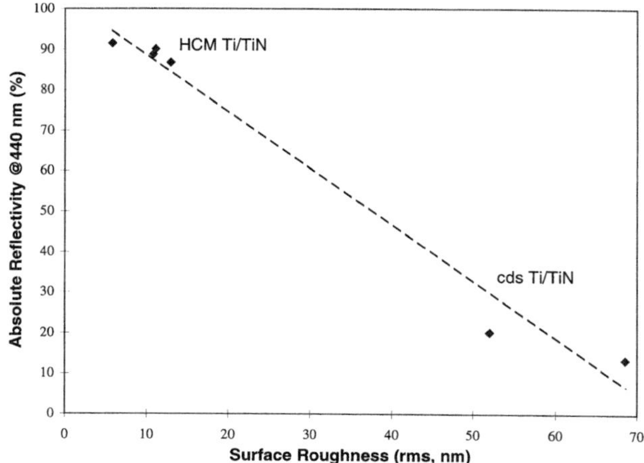

FIG. 25. Dependence of LP Al reflectivity with surface roughness for 7 kÅ of LP Al on HCM and collimated Ti/TiN of various Ti thickness [46], © 1998 IEEE.

Figures 27 and 28 present the cross-sectional SEM views of LP-planarized Al on 0.35-μm aspect ratio 2.5 vias, using HCM Ti as the wetting layer. Except for a 5-min cool down (idle) in the collimated module in between the HCM Ti and LP Al deposition for Fig. 28, the process conditions were identical. The dramatic differences in the LP Al filling as shown in the micrographs can be explained by the sensitivity of the Ti surface to residual gas contamination. In contrast, void-free filling of LP Al

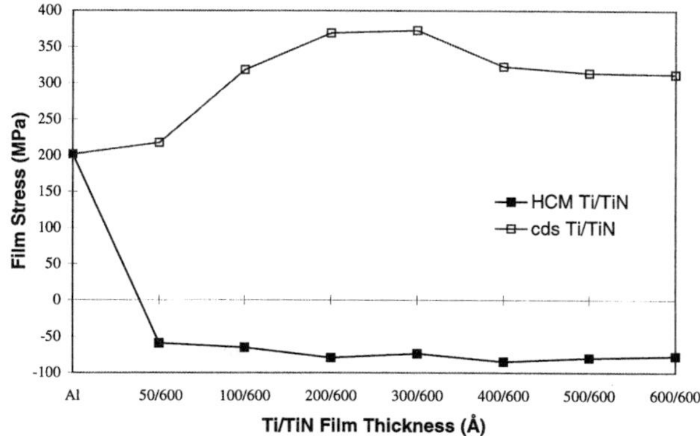

FIG. 26. Effect of Ti underlayer thickness to film stress for 7 kÅ of LP Al on HCM and collimated Ti/TiN [46], © 1998 IEEE.

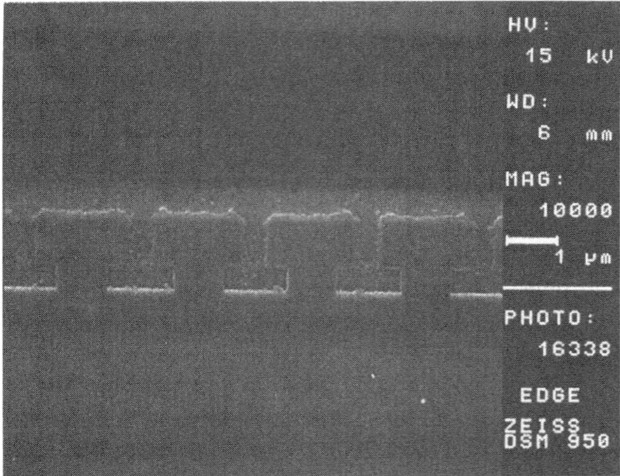

FIG. 27. SEM cross-sectional view of LP Al planarization on 0.35-μm-diameter, aspect ratio 2.5 vias using HCM Ti underlayer (no cool step) [46], © 1998 IEEE.

can be consistently achieved even after prolonged periods of idling using the HCM Ti/TiN underlayer as shown in Fig. 29. Comparing Figures 27 and 29, the HCM Ti/TiN has significantly less $TiAl_x$ formation than does HCM Ti.

The HCM Ti/TiN has been successfully integrated with the LP Al planar-

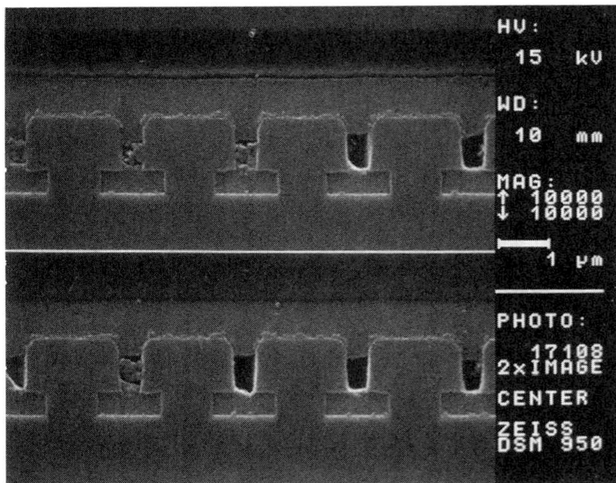

FIG. 28. SEM cross-sectional view of LP Al planarization on 0.35-μm-diameter, aspect ratio 2.5 vias using HCM Ti underlayer (5 min cool) [46], © 1998 IEEE.

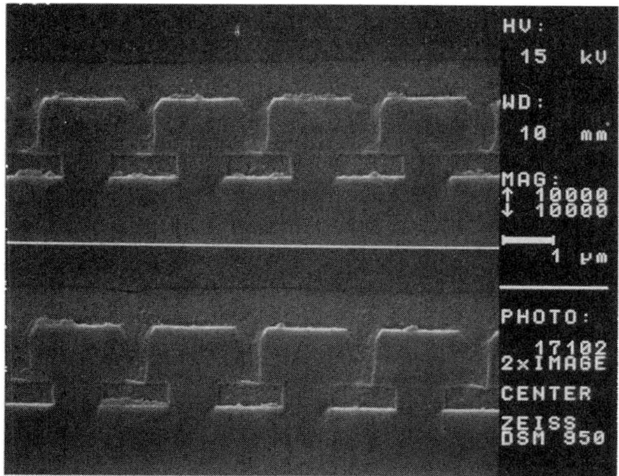

FIG. 29. SEM cross-sectional view of LP Al planarization on 0.35-μm-diameter, aspect ratio 2.5 vias using HCM Ti/TiN underlayer (5-min cool) [46], © 1998 IEEE.

ization process. The new process offers a wide process window, reduced $TiAl_x$ formation, and strong Al(111) grain orientation. In addition, the HCM has much better step coverage than collimated sputtering, making it attractive for filling sub-0.35-μm vias.

B. Al Underlayer for Al Planarization

The filling of dual-damascene structures by conventional PVD Al reflow at elevated temperature is a marginal process for feature sizes smaller than 0.25 μm and aspect ratio >2.8.[49] While the use of the HCM Ti/TiN wetting layer together with LP Al can extend the reflow process to an even higher aspect ratio, the decrease of sidewall Al thickness with the increase in aspect ratio eventually prevents reliable via filling beyond aspect ratio 4. One possible solution is the use of a CVD Al underlayer followed by a hot PVD AlCu reflow. While the CVD/PVD Al approach has been shown to be reliable in filling structures down to 0.175 μm and aspect ratio 4.5,[49] integration of multiple PVD and CVD steps (I-PVD Ti, CVD TiN, CVD Al, and PVD AlCu) in the same tool raises concerns of cross-contamination by the transfer chamber and by the wafer because of the four orders of magnitude differences in process pressures between the PVD and CVD modules. Some degradation in the PVD Al purity is a necessary compromise to avoid excessive throughput penalty for the CVD/PVD integration.

Due to its lower cost of ownership, hardware simplicity, and higher film purity, PVD Al filling is preferred over the CVD/PVD approach provided that the PVD process can be extended to fill higher aspect ratio (>4) vias. Since HCM deposition provides much better step coverage than collimation, it is possible to use HCM Al to replace collimated Al to improve the sidewall coverage in the initial cold Al deposition step. Figures 30a and 30b show the initial results of filling 0.30-μm vias of aspect ratio 4.3 by the LP Al planarization process using the HCM Al and collimated Al underlayer, respectively. Due to slight misalignment in the sample polishing, only the center via was cross-sectioned properly. Unreliable via filling using collimated Al as the underlayer can be clearly seen in Fig. 30b as evident by the void near the bottom of the via because of insufficient sidewall coverage. Due to the much improved sidewall coverage of the HCM Al, reliable filling of the vias can be achieved as shown in Fig. 30a. HCM Ti/TiN was used as the wetting layer for both the HCM and collimated cold Al deposition, whereas LP Al sputtering was used for the final hot Al planarization.

C. Cu Seed Layer and Fill

Copper is an attractive material for next-generation high-performance interconnects because of its low resistivity and resistance to electromigration. Copper has a resistivity of 1.67 $\mu\Omega$-cm, which is approximately 30% less than that of pure aluminum and almost 50% less than that of Al-Cu(0.5–4%)Si(0–1%) alloys, which have been the dominant interconnect material for the past 20 years. As critical dimensions are scaled down to below 0.25 μm, aluminum metalization faces severe reliability problems because of much increased current density. Studies have shown copper to have electromigration lifetimes up to two orders of magnitude greater than those of conventional Al-0.5% Cu lines.[50]

1. Cu Seed Layer for Electroplating

Electroplating offers an attractive alternative deposition method for copper that is not available for tungsten or aluminum. While electroplating is an inexpensive process in principle and has proven to be successful in filling high aspect ratio damascene structures, it is intrinsically a two-step process. A thin seed layer is required prior to the plating step to provide a low-resistance conductor for the plating current and to facilitate the nucleation of the plated copper film. Copper is the preferred seed layer material because of its ideal nucleation characteristics for electroplated copper film and also its high electrical conductivity. On the wafer scale, the thickness of the seed

a

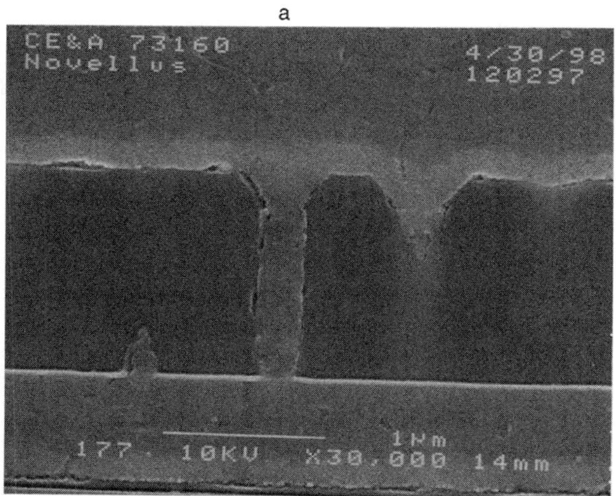

b

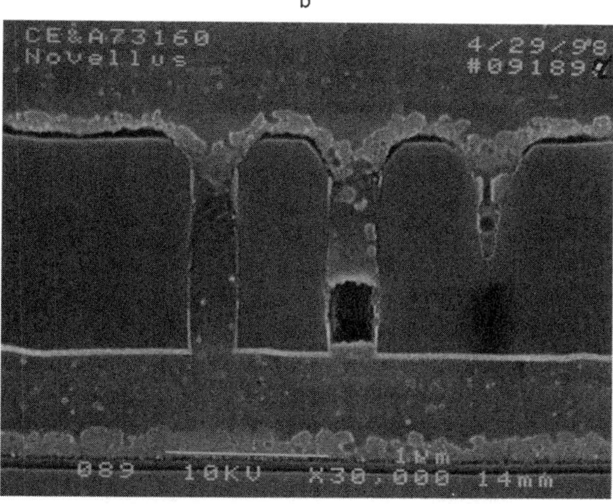

FIG. 30. SEM cross-sectional view of LP Al fill on 0.30-μm-diameter, aspect ratio 4.3 vias using (a) collimated Al and (b) HCM Al, as the cold Al underlayer respectively (courtesy of Dr. M. Rumer).

layer must be sufficient so that the voltage drop from the wafer edge to center does not negatively impact the within-wafer uniformity of the plating process. On a feature size scale, the seed layer carries current from the field level to the bottom of vias and trenches. In addition to the need for sufficient seed layer thickness along the sidewall and bottom of the structure to avoid excessive voltage drop, the seed layer must be continuous and smooth enough to prevent premature closing during plating. In addition, the microstructure of the seed is preferred to be (111) textured so that the plated Cu has a similar texture for good electromigration resistance.

Although PVD copper film has limited step coverage in high aspect ratio vias and trenches, it has been successfully applied to electroplated fill in lower aspect ratio structures. Because of its much better step coverage, HCM copper deposition can extend PVD copper seed layer to much higher aspect ratio structures. Figure 31 shows the step coverage obtained by HCM Cu on 0.20 μm-wide-trenches of aspect ratio 7.4 lined with collimated Ta barrier. The Ta field thickness was intentionally increased to 1000 Å to facilitate step coverage measurement by SEM. Although the trenches are slightly reentrant after the collimated Ta deposition, the HCM Cu provides a nearly conformal, continuous copper seed layer inside the high aspect ratio structure.

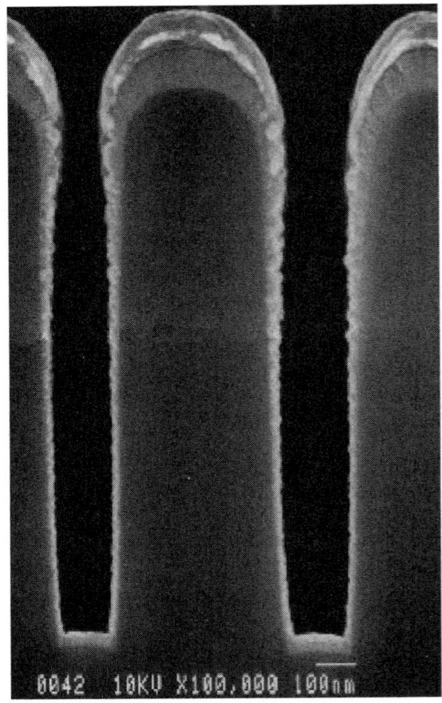

FIG. 31. SEM cross-sectional view of 0.20-μm-wide, aspect ratio 7.4 trenches deposited with 1000 Å HCM Cu seed layer and 1000 Å collimated Ta diffusion barrier (courtesy of Dr. E. Klawuhn).

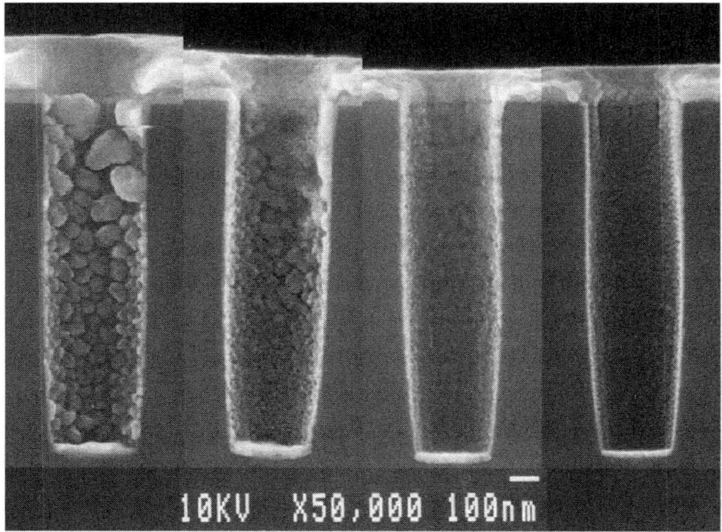

FIG. 32. Cross-sectional SEM micrograph of Cu seed layer deposited by HCM sputtering on 0.35-μm-diameter, aspect ratio 3.6 vias for four different samples of decreasing substrate temperatures ($T_a > T_b > T_c > T_d$).

Another important property of a thin copper film is its tendency to agglomerate at elevated deposition temperatures. Figure 32 shows the grain structure of an HCM copper film inside a 0.35-μm-diameter via of aspect ratio 3.6 for four different samples deposited at decreasing substrate temperatures ($T_a > T_b > T_c > T_d$). As can be seen, the lower the deposition temperature, the smaller the size of the Cu agglomeration. A continuous copper seed layer can only be obtained if the thickness of the copper film exceeds the size of the agglomerated Cu clusters. Rapid grain growth under low-energy plasma bombardment[51] is probably an important factor to agglomeration as can be seen in Fig. 32a in which the size of the agglomerated cluster along the sidewall decreases from the opening to the bottom of the via.

2. *Cu Fill by Cold/Hot Deposition*

I-PVD alone has been shown to be effective in filling low aspect ratio damascene structures. By coupling a permanent magnet ECR source[52] or an ICP source[53,54] to a copper sputtered target, complete filling of trenches of aspect ratio <2 has been reported. Because of the lack of 100% directionality of the ionized deposition flux in these experiments, the top edge of a high aspect ratio structure tends to pinch off and it creates a void before it can be completely filled from the bottom up. An increase of bias voltage and thus the incident ion energy results in an increase in the amount of reflected metal ions and resputtering. While this is beneficial to eliminate the formation of overhang and columnar sidewall microstructure, and the filling of seams in the corners, it also results in the growth of wall deposit which can be detrimental beyond a threshold bias voltage (typically -30 V). In an extreme case in which the ionized flux approached 100%, feasibility of filling via structure of aspect ratio up to 4.2 has been demonstrated[55] using a two-step bias scheme with high bias voltages ($\geqslant 100$ V). High bias voltage, however, has raised the concerns of top corner beveling, charging damage, high film resistivity, and argon atom incorporation. Therefore, I-PVD alone is not likely to fill and planarize structures of aspect ratio much greater than 3 in VLSI metalization.

The combination of long-throw sputtering followed by heat treatment[56] or conventional sputtering followed by high-pressure reflow[57] have shown to be capable of filling submicron structures of aspect ratios up to 3 and 5, respectively. Due to the long process time (typically 5 min) and high reflow temperature ($\geqslant 440$°C), however, the reflow process is not deemed cost-effective nor compatible with low dielectric constant materials.

With its much improved directionality and increased surface mobility due to ion bombardment, the combination of I-PVD with a two-step (cold/hot) deposition process may be an attractive solution in the filling of high aspect ratio (>3) damascene structures at moderate deposition temperatures (<400°C). The feasibility of using the HCM copper deposition together with the two-step process to fill high aspect ratio trenches and vias are shown in Figs. 33 and 34, respectively. The substrate was clamped on an heater stage with backside argon cooling. Although the stage was left at room temperature for this specific deposition, the temperature of the wafer drifted higher (probably to ~ 200°C near the end of deposition) due to the high-density plasma heating and the limited cooling capability of the backside argon gas. As can be seen in Fig. 33, complete filling of 0.35-μm-wide trenches of aspect ratios 1.5 can be easily achieved with HCM Cu deposition. The filling of vias of aspect ratio 2.7 were sporadic, with completely filled vias next to empty ones as shown in Fig. 34. The variation of via filling can probably be explained by the agglomeration of the copper seed layer on the sidewall as discussed in the previous section. Since the reflow process depends on the quality and continuity of the copper seed layer, a cold step for the formation of an agglomeration-free seed layer followed by a hot deposition step for the reflow of the copper will be required to alleviate the inconsistent via filling as shown in Fig. 34.

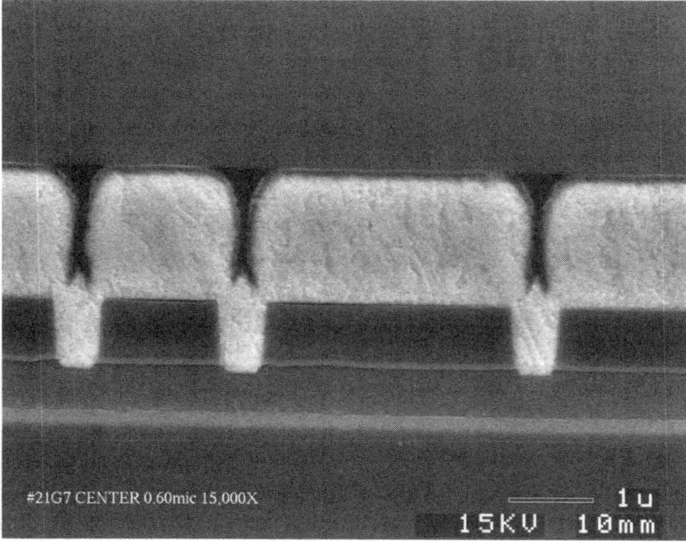

FIG. 33. Cross-sectional SEM micrograph of 0.35-μm-diameter, aspect ratio 1.5 trenches filled by HCM Cu deposition.

FIG. 34. Cross-sectional SEM micrograph of 0.35-μm-diameter, aspect ratio 2.7 vias filled by HCM Cu deposition.

D. Ta/TaN DIFFUSION BARRIERS

Tantalum-based diffusion barriers consisting of elemental tantalum, tantalum with light amounts of nitrogen doping, or a full tantalum nitride compound are the materials of choice for the initial manufacturing of copper interconnects.[58] While tantalum is an attractive material because of its high melting point and its immiscibility with copper,[59] doping the tantalum film with a few percent of nitrogen has been shown to be effective in blocking the grain boundary diffusion paths and further enhances its diffusion barrier property. Due to the nonunity sticking coefficient and strong angle-dependent reflection coefficient of tantalum,[60] collimated sputtering of tantalum or tantalum nitride by reactive sputtering has been shown to provide adequate step coverage for high aspect ratio ($\leqslant 5$) trenches. The step coverage of PVD tantalum-based diffusion barriers can be further extended by employing I-PVD techniques making it an attractive diffusion barrier deposition technique even for very high aspect ratio structures.

Figures 35a and 35b show the cross-sectional TEM views of HCM Ta film deposited without applied wafer bias on 0.27-μm-diameter, aspect ratio 4.3 contacts. Again, near conformal coating is achieved even around the bottom corner where the contact is slightly reentrant. The bottom and minimum sidewall coverages of HCM Ta are 42 and 17% respectively, which are much better than those deposited by collimated sputtering. As expected, the step coverage of HCM Ta is significantly higher than that of HCM Ti (Fig. 21) due to the nonunity sticking coefficient of Ta. The diffusion barrier property of HCM Ta can be further enhanced with the addition of nitrogen, as the grain size of TaN_x decreases with increasing nitrogen concentration.[61]

V. Other Applications

Because of its extremely high plasma density, hardware simplicity, and scaleability, the HCM can also be used in applications other than I-PVD. In this section, two possible applications in which the HCM may be used are briefly discussed.

The HCM can easily be converted into an intense ion source by placing an ion extraction aperture or a slit near the waist of the emitted plasma beam. With proper magnetic design, the plasma beam from a cylindrical or rectangular cathode can be focused to the shape of the extraction aperture or slit respectively to maximize the ion extraction efficiency. The quality of the extracted ion beam is expected to be high, because of the cusp magnetic mirror and the avoidance of the cross-field instability as in the IRMA ion

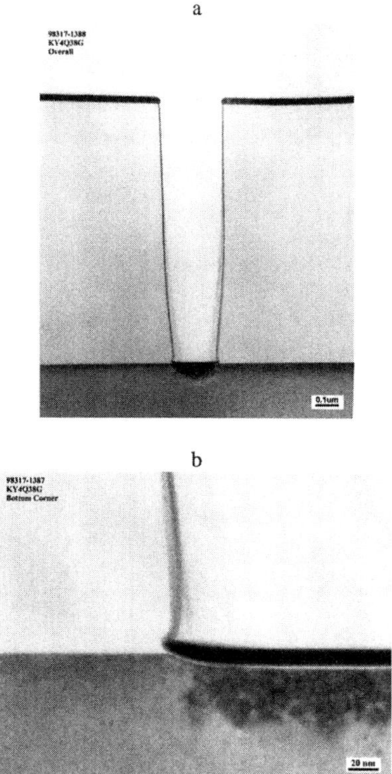

FIG. 35. TEM cross-sectional views of a 300 Å Ta film deposited by HCM sputtering on 0.32-μm-diameter, aspect ratio 3.8 contacts (a) overall and (b) bottom corner (courtesy of Dr. E. Klawuhn).

source. The HCM ion source is uniquely suitable for generating an ion beam from solid materials. For ion implantation application, solid target materials, such as boron, boron carbide, metal borides, and boron silicide (BSi_x) can be used instead of toxic or highly flammable boron-containing gases such as BF_3 or B_2H_6 in the generation of boron or boron-containing ions. Boron silicide in particular is a very promising material for the generation of BSi_x^+ for extremely shallow junction ion implantation. Because of its small size, hardware simplicity (no filament or moving parts), and low maintenance, the HCM ion source should be a strong candidate for solid ion source for ion implantation application.

Due to its extremely high-density plasma, the HCM has intense radiation in both the vacuum ultraviolet (VUV) and ultraviolet (UV) spectra. With proper design, the HCM can be converted from a deposition source into a highly focused VUV and UV light source. By operating the HCM in sub-mTorr pressure to increase the electron temperature to $\gg 10$ eV, efficient ionization and excitation of the target and feed gas atoms can be achieved to significantly enhance the amount of radiation in the VUV and UV spectra from the plasma. With proper selection of the target materials and the feed gas, many radiation lines will be accessible with the HCM light source.

VI. Conclusions

The HCM has been shown to be a very successful high-density plasma source for I-PVD application. It has been proven to be reliable, scaleable, versatile, and suitable for the deposition of a large variety of materials with minimal modification of the source. Its ability to operate in the nonpoisoned mode for reactive sputtering has been extremely valuable in the deposition of binary compound films. The characteristics of HCM deposited films can be dramatically different from those by conventional PVD. More research is needed to understand the physics of metal ion deposition under high-flux, low-energy ion bombardment as evident by the extremely low-resistivity TiN and the near-conformal step coverage of high *AR* features by HCM deposition. Because of its simplicity and hardware compatibility with existing PVD deposition systems, the HCM promises significant advantages in comparison to other I-PVD approaches.

After more than 10 years since its first conception, the HCM has finally passed all the hurdles in research and development and has reached the commercialization stage. Figure 36 shows a commercial HCM source for 200-mm-diameter wafer metallization offered by Novellus Systems. It is expected that the HCM technology will continue to evolve and improve its capability to fulfill the metallization needs of future generations of ULSI devices for both 200- and 300-mm-diameter wafers.

Acknowledgments

I am in debt to many current and former coworkers in Novellus and Varian for their contributions and many helpful discussions. In particular, I thank Dr. Q. Lu for her help in collecting and analyzing many of the data presented in this chapter, Dr. L. M. Tam for his collaboration on HCM Ti/TiN deposition, Dr. C. D'Couto for his data on HCM Ti/TiN barrier deposition, Dr. E. Klawuhn for his data on HCM copper seed layer and tantalum barrier deposition, Dr. M. Rumer for his data on HCM aluminum

FIG. 36. A commercial HCM sputtering source for 200-mm-diameter wafer deposition offered by Novellus Systems.

deposition, Dr. M. Biberger for his support, and Dr. R. Powell for his support and critical review of the manuscript. Acknowledgment also is due to Dr. J. Feinstein, Dr. L. D. Hartsough, A. L. Nordquist, and Dr. K. Ashtiani for their many useful discussions. I also express my appreciation to my former coworkers, Drs. J. C. Helmer and R. L. Anderson, for introducing me to the HCM concept, and Dr. G. J. Reynolds for his support which accelerate the HCM development. Finally and most important, I express my deepest appreciation to my wife, Houchin, for her support, encouragement, and tolerance to allow me to write this chapter.

References

1. J. Drewrey *et. al.*, Proceedings of the 14th VLSI Multilevel Interconnect Conference, p. 274 (1997).
2. G. W. White, U.S. Patent No. **4420386**, Dec. 13, 1983.
3. S.-N. Mei, T.-M. Lu, and S. Robert, *IEEE Electron Device Lett.* **EDL-8**, 503 (1987).
4. J. R. Morley and H.R. Smith, *J. Vac. Sci. Technol.* **9**, 1377 (1972).
5. J. Uramoto, *J. Vac. Soc. Jap.* **25**, 660 (1982).
6. S. Komiya and K. Tsuruoka, Proceedings of the 6th International Vacuum Congr., p. 415 (1974).
7. H. Makino, M. Tanaka, and K. Awai, Proceedings of the 15th VLSI Multilevel Interconnect Conference, p. 75 (1998).

8. C. V. Deshpandy and R. F. Bunshah, in *Thin Film Processes II* (J. L. Vossen and W. Kern Eds.), p. 79, Academic Press, San Diego, 1991.
9. M. Yamashita, *J. Vac. Sci. Technol. A* **7**, 151 (1989).
10. P. W. Kidd, U.S. Patent No. **4925542**, May 15, 1990.
11. M. S. Barnes, J. C. Forster, and J.H. Keller, U.S. Patent No. **5178739**, January 12, 1993.
12. D. B. Fraser, in *Thin Film Processes* (J. L. Vossen and W. Kern Eds.), p. 115, Academic Press, New York, 1978.
13. S. M. Rossnagel and H. R. Kaufman, *J. Vac. Sci. Technol. A* **4**, 1822 (1986).
14. J. A. Thornton, *J. Vac. Sci. Technol.* **15**, 171 (1978).
15. J. C. Helmer and K. J. Doniger, U.S. Patent No. **4774437**, September 27, 1988.
16. J. C. Helmer, personal communication.
17. G. R. Lavering, Varian internal design drawing, September 1987.
18. V. L. Hedgcoth, U.S. Patent No. **5073245**, December 17, 1991.
19. J. A. Thornton and A. S. Penfold, in *Thin Film Processes* (J. L. Vossen and W. Kern Eds.), p. 76, Academic Press, San Diego, 1978.
20. V. A. Gruzdev, Yu. E. Kreindel', and O. E. Troyan, *Sov. Phys. Tech. Phys.*, **25**, 1228 (1980).
21. N. Kumar, J. Pourrezaei, and M. Ihsan, *J. Vac. Sci. Technol. A* **6**, 1772 (1988).
22. Y. Hoshi, N. Terada, M. Naoe, and S. Yamanaka, *IEEE Trans. Magnet.* **MAG-17**, 3432 (1981).
23. A. P. Semenov, *Sov. Phys. Tech. Phys.* **32**, 109 (1987).
24. V. Miljevic, *Rev. Sci. Instrum.* **55**, 121 (1984).
25. R. D. Rust, U.S. Patent No. **4915805**, April 10, 1990.
26. H. Kawasaki, T. Nakashima, and H. Fujiyama, *Mater Sci. Eng. A* **140**, 682 (1991).
27. J. J. Cuomo and H. R. Kaufman, U.S. Patent No. **4588490**, May 13, 1986.
28. J. C. Helmer, K. F. Lai, and R. L. Anderson, U.S. Patent No. **5482611**, January 9, 1996.
29. F. F. Chen, *Introduction to Plasma Physics*, Plenum, New York, 1974.
30. J. Hopwood and F. Qian, *J. Appl. Phys.* **78**, 758 (1995).
31. M. A. Vidal and R. Asomoza, *J. Appl. Phys.* **67,** 477 (1990).
32. J. C. Helmer, Varian internal memo (1987).
33. R. K. Waits, in *Thin Film Processes* (J. L. Vossen and W. Kern Eds.), p. 131, Academic Press, San Diego, 1978.
34. S. M. Rossnagel, *J. Vac. Sci. Technol. A* **6**, 19 (1988).
35. J. L. Vossen and J.J. Cuomo, in *Thin Film Processes* (J. L. Vossen and W. Kern Eds.), p. 11, Academic Press, San Diego, 1978.
36. M. Vukovic and K. F. Lai, presented in Gaseous Electronic Conference (1997).
37. J. B. Hasted, *Int. J. Mass Spect. Ion Phys.* **16**, 3 (1975).
38. D. C. Smith, Schumacher CVD Symposium, p. 15 (1997).
39. G.A. Dixit *et. al.*, IEDM 1996 (1996).
40. M. Biberger, S. Jackson, M. Rumer, and G. Tkach, *Semicon Korea 95*, p. 89 (1995).
41. K. F. Lai *et. al.*, Proceedings of the 14th VLSI Multilevel Interconnect Conference, p. 234, (1997).
42. B. O. Johansson, J.-E. Sundgren, and J. E. Greene., *J. Vac. Sci. Technol. A* **3**, 303 (1985).
43. H. P. Klug and L. E. Alexander, *X-ray Diffraction Procedures*, Wiley, New York, 1954.
44. M. Biberger *et. al.*, Proceedings of the 4th International Symposium on Sputtering & Plasma Processes, p. 299 (1997).
45. I. S. Park *et. al.*, Proceedings of the 12th VLSI Multilevel Interconnect Conference, p. 45 (1995).
46. K. F. Lai, L. M. Tam, and Q. Lu, Proceedings of the International Interconnect Technology Conference, p. 292, IEEE (1998).

47. J. Su *et. al.*, Proceedings of the 15th VLSI Multilevel Interconnect Conference, p. 124 (1998).
48. K. Hinode, S. Kondo, and O. Deguchi, *J. Vac. Sci. Technol. B* **14**, 687 (1996).
49. R. Iggulden *et. al.*, Proceedings of the 15th VLSI Multilevel Interconnect Conference, p. 19 (1998).
50. D. Pramanik, *MRS Bulletin*, 57 (1995).
51. M. D. Naeem, S. M. Rossnagel, *J. Vac. Sci. Technol. B* **13**, 209 (1995).
52. S. M. Gorbatkin *et. al.*, *J. Vac. Sci. Technol. B* **14**, 1853 (1996).
53. C. A. Nichols, S. M. Rossnagel, and S. Hamaguchi, *J. Vac. Sci. Technol. B* **14**, 3270 (1996).
54. P. F. Cheng, S. M. Rossnagel, and D. N. Ruzic, *J. Vac. Sci. Technol. B* **13**, 203 (1995).
55. W. M. Holber *et. al.*, *J. Vac. Sci. Technol. A* **11**, 2903 (1993).
56. T. Saito *et. al.*, Proceedings of the International Interconnect Technology Conference, p. 160, IEEE (1998).
57. K. Maekawa *et. al.*, Proceedings of the International Interconnect Technology Conference, p. 169, IEEE (1998).
58. P. Singer, *Semiconductor Int.* **21**, 90 (1998).
59. M. Takeyama, A. Noya, T. Sase, A. Ohta, and K. Sasaki, *J. Vac. Sci. Technol. B* **14**, 674 (1996).
60. S. M. Rossnagel *et. al.*, *J. Vac. Sci. Technol. B* **14**, 1819 (1996).
61. X. Sun, E. Kolawa, J.-S. Chen, J. S. Reid, and M.-A. Nicolet, *Thin Solid Films* **236**, 347 (1993).

Applications and Properties of Ionized Physical Vapor Deposition Films

JOHN FORSTER

Applied Materials, Santa Clara, California

I. Introduction

By now, readers of the previous chapters should have a good background in the basic physics underlying ionized physical vapor deposition (I-PVD) or ionized metal deposition. The evolution of a plasma source from laboratory curiosity to an industrial tool requires several prerequisites. One of the most important is the ability of the source to perform applications better than other available technologies. This chapter will focus on the applications of the I-PVD source in the microelectronics industry.

Research into I-PVD sources has been ongoing for many years. It is only recently, however, that the microelectronics industry has progressed to the point where the introduction of I-PVD sources becomes attractive from an economical point of view. The creation of integrated circuits (ICs) requires the ability to deposit conducting layers onto the silicon wafer and, very often, the ability to deposit conducting material into very high aspect ratio (i.e., deep and narrow) features. I-PVD sources are able to deposit metal into small, high aspect ratio features much more readily than their older counterparts, such as collimated PVD or standard PVD. Thus, it is expected

Vol. 27
ISBN 0-12-533027-8

ISSN 1079-4050/00 $30.00

that I-PVD technology will become established as the industry moves beyond 0.35-μm technology (equivalent to ~64 Mbit DRAM technology). The I-PVD source has also generated interest because it offers more input variables ("knobs") for controlling film properties of the deposited metal.

This chapter will describe some of the applications of I-PVD technology and the properties of films deposited with I-PVD technology. The chapter will start with a brief overview of the process flow of ICs during the metallization process. A discussion of the process results achieved with I-PVD sources will follow, along with descriptions of specific applications. It should be noted that this review of applications is not exhaustive. There are probably references in the literature which we overlooked. Therefore, we apologize in advance for omissions. Also, apologies are offered for referring so often to conference proceedings, which may be difficult for some readers to obtain. This is an unfortunate consequence of the very recent introduction of commercial I-PVD sources.

II. Metallization for Integrated Circuits

There are several excellent texts on the process steps involved in semiconductor manufacturing.[1,2] Only a brief overview of the processes involving metal deposition can be given here. The fabrication of ICs requires the ability to deposit thin films of metals onto the silicon wafer or substrate. These processing steps occur after the transistors have been built and are sometimes referred to as back-end-of-line processing steps. The metal serves as the electrical connections between the transistors. The complex interconnections between the transistors in modern ICs require the use of several layers of metal wiring. The typical layout of a three-layer metallization scheme is shown in Fig. 1. At the bottom of the figure is the transistor. On top of the transistor is the first interlevel dielectric (ILD), which is typically silicon dioxide (SiO_2). Above the first-level ILD is the first layer of metal (often referred to as "metal 1"). The metal in the layers is patterned into lines to produce the electrical pathways between the circuit elements of the IC. A hole filled with metal (the "contact") serves as a connection between the first metal layer and the transistor. Above metal 1 are more ILD layers and metal layers, with metal-filled holes ("vias") serving as connections between the layers. The astute reader may have deduced that manufacturing the metallization scheme depicted in Fig. 1 puts two basic requirements on metal deposition technology. It requires the ability to deposit metal into the contact and via holes in the ILD. As device and, subsequently, hole dimensions shrink, it becomes increasingly difficult to direct metal into the holes. Producing the metal layers also requires the ability to deposit metal

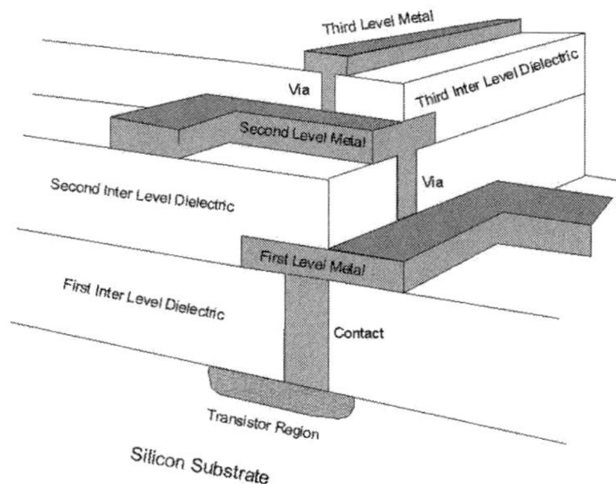

FIG. 1. Schematic of metallization in an IC.

in a flat, even, and bump-free manner, a process referred to as "planarization." Denboer[3] provides pictures of the metallization structure of actual ICs.

Currently, there are two schemes for producing the metal lines, schematically depicted in Fig. 2. The traditional approach involves depositing the metal as a flat blanket film and then using a metal etch step to form the lines. Note that this approach requires the subsequent ILD deposition step to fill the gaps between the line. This method of producing metal lines places stringent demands on both the metal etch and the ILD deposition process. A newer scheme involves using a dielectric etch process to create trenches in the dielectric. The metal lines are created by depositing metal into the trenches. In this scheme, called "damascene," the process that fills the lines must be followed by a planarization step that removes metal from the areas between the lines. The damascene process places stringent demands on the oxide etch process. In addition to the single-damascene approach shown in Fig. 2, there is also a "double-damascene" approach in which the trenches and vias are both etched prior to metal deposition and filled during a single metal deposition step. Wolf[1] and Murarke[2] provide more information on the required process flows.

Currently, the metal predominantly used for the metal layers is aluminum (Al). As device speed increases, the need to reduce RC delays of the interconnects will force IC manufacturers to use copper (Cu) for interconnects. Also, as line dimensions shrink, electromigration-induced effects will become more prevalent, and Cu is more resistant to electromigration than Al. The introduction of the first commercially available circuits with Cu interconnects almost coincided with the publication of this book.

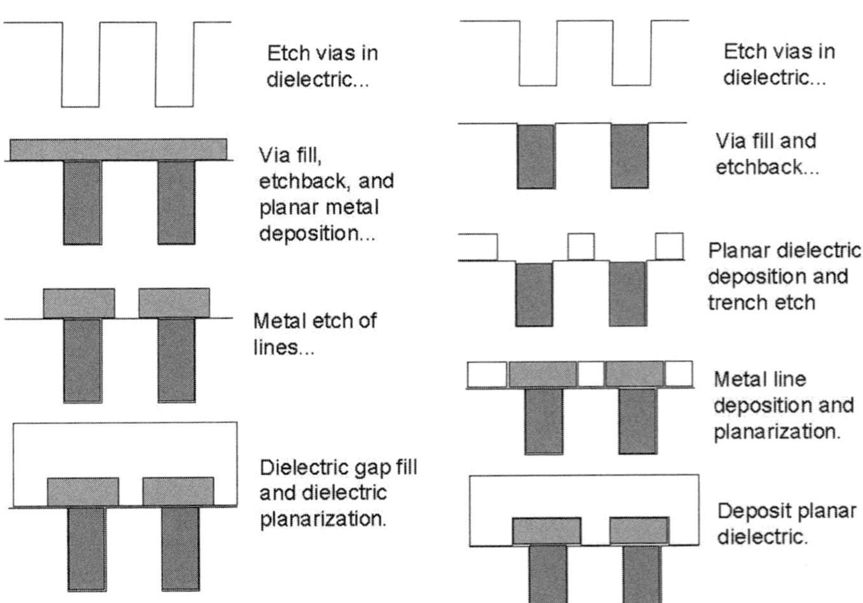

FIG. 2. Comparison of metal etch and damascene metallization schemes: (a) metal etch scheme and (b) single damascene.

The holes between the various layers of metal are called contacts if they are between metal 1 and the transistor, and they are called vias if they are between two metal layers. Once the holes are filled with a metal, they are sometimes called "plugs." The metals predominantly used for filling contacts and vias are Al and tungsten (W), with imminent introduction of Cu-filled contacts and vias.

Other metals and conducting compounds, such as titanium (Ti), tantalum (Ta), titanium nitride (TiN), and tantalum nitride (TaN), are used as adhesion promoters and as diffusion barriers. Figure 3 shows three different applications of Ti and TiN. Figure 3a shows a contact. The first metal to be deposited is Ti. Note that the Ti layer is in direct contact with the Si of the underlying transistor. In order to reduce the contact resistance, the structure will be subsequently annealed at a high temperature ($700^\circ\mathrm{C} < T < 900^\circ\mathrm{C}$), causing formation of $TiSi_2$ at the interface between the Ti and the Si. After deposition of Ti, the contact must be filled with either W or Al. Figure 3b shows how Ti and TiN layers are used in forming a W-filled contact or via. Again, Ti is the first layer deposited. The subsequent layers do not adhere well to the ILD, but they adhere well to Ti; thus, Ti serves as an adhesion

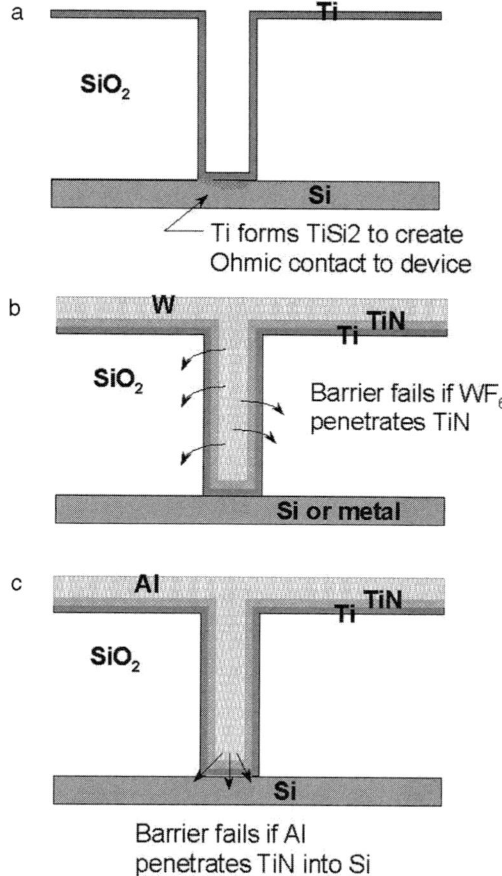

FIG. 3. Typical contact and via applications: (a) contact silicide, (b) tungsten fill of contact or via, and (c) Al fill of contact or via.

promoter. Deposition of W is commonly performed by chemical vapor deposition (CVD) from WF_6. The WF_6 thermally decomposes at the substrate; the W adheres to the substrate, and the fluorine desorbs into a gaseous state. The WF_6 is extremely corrosive toward Ti and the ILD material. A thin layer of TiN is used to protect the Ti and the ILD from the corrosive W-CVD environment. Figure 3c shows an Al-filled contact. Again, Ti is used to form a good ohmic contact to the underlying Si. Al reacts readily with Si, and any penetration of Al into the Si would render the transistor useless. Therefore, a layer of TiN is placed between the Al and Si as a diffusion barrier. Finally, a thin layer of Ti (not shown) is deposited on the TiN to promote adhesion of the Al. This thin Ti layer is referred to as a "wetting layer." If Al plugs are used at the via level, then there is no issue of Al penetrating the barrier, so either the previously described trilayer of Ti/TiN/Ti or just a single Ti layer can be used to line the via prior to Al planarization. Wolf,[1] Wittmer,[4] and Nicolet[5] review diffusion barriers for microelectronic applications. Ouellet *et al.*[6] describe the interaction between the liner/barrier layers and the Al layer in a contact metallization.

For Cu-based interconnects, the barrier material would be Ta or TaN. In many cases, the contacts would still be filled with a W-plug process, whereas Cu would be used at the via levels. The barrier requirements for Cu-based interconnects are more severe than those for Al-based interconnects. Copper can diffuse rapidly through Si and SiO_2. Thus, the Cu must be completely encapsulated by a barrier to prevent Cu diffusion. Either electroplating or CVD processes have been demonstrated for via-level Cu fill. Cu fill using electroplating will often require deposition of a thin Cu seed layer prior to electroplating.

The preceding overview was very brief, and the reader can see the references for a better explanation of the process steps required to produce an integrated circuit.[1,2].

III. Characteristics of Film Deposition Using an I-PVD Source

The process engineer is not concerned with the inner workings of a deposition source. Of interest are only the properties of the deposited film. Additionally, in a manufacturing environment, the source must be able to deposit films repeatedly with identical properties, regardless of time of day, phase of the moon, or past history of the source. The most fundamental process characteristics are deposition rate, uniformity of the film over the wafer, and step coverage. These are properties that are fairly easy to measure. In some situations, there may be a desire to control film microstructure because it may have consequences for final device performance.

A. Deposition Rate and Uniformity

The deposition rate in a deposition source is important because it will determine the time to process a wafer and the number of wafers that can be processed in a given amount of time (throughput). While the throughput may only be of slight interest to a R&D scientist, it is of primary interest to manufacturing managers and is of great importance for commercialization of a deposition source. Deposition rate requirements vary with the application. A rule of thumb is that no deposition process should last much longer than a minute. Liner and barrier layers (e.g., Ti, TiN, Ta, and TaN) require the deposition of thin films (< 1000 Å) and thus require fairly low deposition rates (< 1000 Å/min). Direct fill of plugs or deposition for planarization with Al or Cu requires deposition of thick metal films (> 5000 Å) and thus requires high deposition rates (> 5000 Å/min).

The deposition rate in an I-PVD sources depends mainly on target power. The target power supplies the energy for sputtering the target. As explained in previous chapters, the sputtering rate is given by the product of ion current and sputter yield, the latter being a function of ion energy. Most PVD sources (standard or ionized) operate in a regime in which the current is only weakly dependent on target power. Thus, any increase in target power goes mainly into increasing target voltage, thereby increasing ion bombardment energy and thereby increasing the sputter yield.

The deposition rate will also depend on pressure. Increasing pressure leads to increased gas scattering of sputtered metal. The deposition rate on the substrate decreases, whereas deposition rate onto the sidewalls of the source and redeposition onto the target increase.

The effect of radio frequency (RF) inductive power on deposition rate depends on the source geometry and design. If the source is driven with an internal coil, then it is possible that increasing the RF inductive power will lead to enhanced coil sputtering, which will lead to an increase in deposition rate.

The uniformity of deposition is important when integrating a given metal deposition process into the process flow of constructing a working device. Nonuniform deposition will place additional burdens on subsequent processes and can also lead to variations in device performance across a wafer, which can negatively impact final yield of the manufacturing operation.

The uniformity of deposition will depend on source design. It is well-known that proper design of the magnet which generates the magnetron discharge in a standard PVD source is essential to obtain good uniformity.[7] The spatial distribution of deposition on the wafer depends on the spatial distribution of the target erosion. Ballistic and Monte Carlo models can be used to predict the uniformity of deposition for standard and collimated PVD sources.[8,9] Modeling of deposition uniformity in an I-PVD source may require fluid models, or a Monte Carlo/fluid hybrid model, due to the high pressures employed in I-PVD sources.[10,11]

The deposition uniformity in an I-PVD source employing an internal coil will also depend on coil sputtering.[12] Most designs of such sources employ coils with a diameter larger than the wafer diameter. Thus, deposition from the coil will tend to be thicker at the edge than at the center. If the target is not much larger than the wafer, and if it is eroded uniformly, then the deposition due to target sputtering will be thicker in the center than at the edge.[10] Figure 4 shows the sheet resistance uniformity of Ti deposited with an I-PVD source as a function of the ratio between RF induction coil power and DC target power.[13] The figure shows how the uniformity is affected by the amount of coil sputtering relative to the amount of target sputtering. The details of the graph will strongly depend on the target and coil geometry and the magnet design.

The uniformity in an I-PVD source employing an external coil and a Faraday shield was measured with an accumulating Langmuir probe.[14–16] The authors found that the uniformity of the neutral Al followed a profile that was consistent with a simple diffusion model. The radial distribution of the Al ion density does not follow a simple model, but the fit of the nonuniformity of the Al ionization fraction to the model was surprisingly good. The simple diffusion model suggests that good uniformities of deposition rate will require a Faraday shield much larger than the target, suggesting a scalability issue for larger wafer sizes for sources utilizing a Faraday shield and an external coil.

Modern magnetron sources can deposit azimuthally very uniform films. In an I-PVD source with an internal coil, there may be azimuthal asymmetries due to the coil. The coil has two ends, and there will an RF voltage drop along the coil. The RF voltage variation along the coil will cause an azimuthal asymmetry in coil sputtering. The details of operating a coil immersed in a plasma have been discussed by several authors.[17–19] Also, several authors have suggested methods to make the RF voltage distribution along the coil more uniform.[20,21]

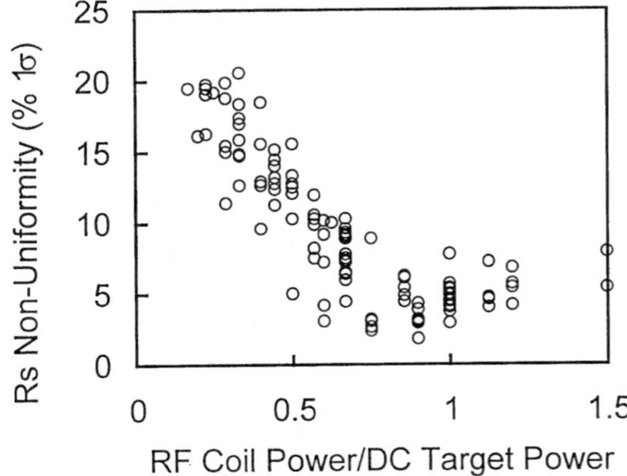

FIG. 4. Uniformity of Ti as a function of the ratio between RF coil power and DC target power for an I-PVD source with an internal coil (from Forster *et al.*[13]).

B. STEP COVERAGE

Before discussing the step coverage requirements, a few terms will be defined. Referring to a schematic depiction of a contact or via hole (Fig. 5), the field thickness is the thickness of material over the flat part of the substrate. The bottom coverage is the ratio between the thickness at the bottom of the hole and the field thickness. The sidewall coverage is the ratio between the thickness of material on the sidewall of the hole and the field thickness. The overhang is the material that causes constriction of the hole opening. Faceting is the occurrence of a sloping sidewall near the top of the hole that can occur in conjunction with overhang.

The requirements for step coverage vary with application. Liner and barrier application usually require a moderate bottom coverage, on the order of 30–70%. Continuous sidewall coverage is usually required to ensure proper protection in the case of a barrier application. For example, the step coverage of Ti at the contact level will directly influence the formation of $TiSi_2$, and it will thereby influence the contact resistance and reverse leakage current of the contact. Other applications, such as Ti and TiN adhesion layers, or a Cu seed layer for subsequent electroplating, require continuous sidewall coverage. The step coverage required for direct-

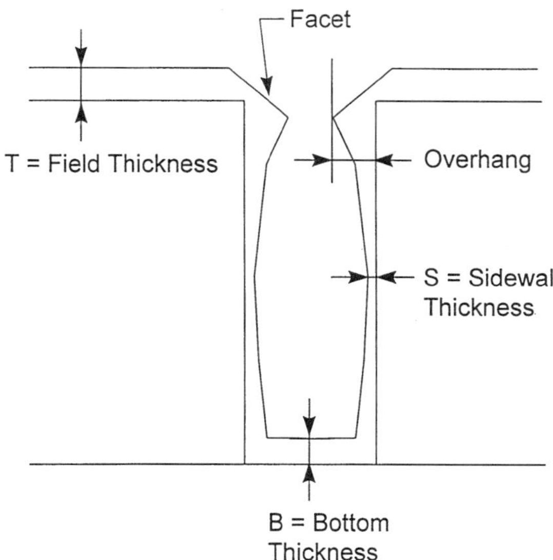

FIG. 5. Typical profile of a via or trench filled by I-PVD. Included are definitions of various terms used in the text.

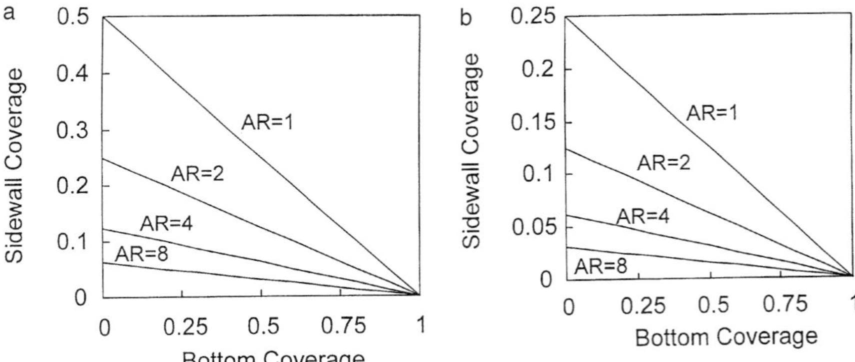

FIG. 6. Theoretical trade-off between sidewall coverage and bottom coverage for various aspect ratio features. Calculation assumes unity sticking coefficient, no resputtering, and uniform sidewall thickness. Features are (a) cylindrical via and (b) rectangular trench.

fill applications is very stringent, $>90\%$, to prevent closing of the hole prior to completion of fill. The amount of tolerable overhang or faceting will also depend on the application, but less overhang will ease the requirements on the subsequent fill process.

For PVD sources, there is a theoretical maximum step coverage for a given hole geometry. In the absence of resputtering, and for a unity sticking coefficient, the total amount of material deposited on the sidewalls and bottom of the hole must equal the volume determined by the opening of the hole and the thickness of the material deposited in the field. Figure 6 shows the trade-off between sidewall coverage and bottom coverage as a function of aspect ratio. There are two graphs, one for holes (Fig. 6a) and the other for trenches (Fig. 6b). The curves on the graph assume no resputtering, a unity sticking coefficient, and uniform sidewall thickness along the length of the feature. Under these assumptions the bottom coverage cannot exceed 100%. A more detailed explanation of the physical mechanisms occurring at the wafer surface during ionized deposition is presented in Refs. 22–24.

1. *Bottom Coverage*

There does not exist a predictive relation between source parameters and bottom coverage. It should be possible to use a plasma simulation to model the metal and ion fluxes in an I-PVD source and then to feed the results to a profile simulator to obtain step coverage as a function of process parameters.[25] Bottom coverage is determined by the directionality of incoming material. In an I-PVD source, the directionality of the incoming material is determined by the ratio between ionized and neutral material and the ion angular distribution function. The latter is influenced by the sheath bias at the substrate and by the bias frequency in the case of an AC bias. It is possible for material deposited into the via to be resputtered by incoming ions, thereby reducing bottom coverage. This effect will increase with increasing bias voltage. As outlined in previous chapters, the ionization ratio of the sputtered metal increases with increasing RF coil power, increasing pressure, and decreasing DC target power.

The bottom coverage has been experimentally shown to follow trends in ionization ratio and bias voltage.[13,23,26,27] Figures 7a–7d show how the bottom coverage of Ti varies as a function of the target power, RF coil power, pressure, and bias voltage. The structure size was $0.35 \times 1.2\ \mu m$, and the bias frequency was 13.56 MHz. The data were obtained from an experimental matrix in which all variables were varied, leading to an apparent large scatter. As expected, the bottom coverage increases with increasing bias voltage, RF coil power, and pressure, and it decreases with increasing target power. The data indicate that bias voltage has the most significant effect on bottom coverage. An ad hoc model of bottom coverage as a function of source parameters, based on experimental results, is shown in Fig. 8. Bottom coverage scales surprisingly well with the function (RF coil power/DC target power) × pressure × bias voltage, despite its simplicity and complete lack of physical basis.

The results for Cu also follow the expected trends for ionization ratio.[23,26,27.] The bottom coverage data reported in Ref. 26 (Fig. 9) were measured in trenches with aspect ratios ranging from 0.8:1 to 1.5:1. The bottom coverage increases with increasing RF coil power and decreasing DC target power. The authors use the bottom coverage as a diagnostic to calculate the ionization ratio of the Cu, which approaches 85%. In Cheng *et al.*,[23] the bottom coverage was measured in 1.1:1 aspect ratio trenches. The bottom coverage increased with increasing RF coil power, increasing pressure, and decreasing DC target power. It did not depend on bias voltage; this may be due to the low aspect ratio (<2:1) of the features used in that work. Alternatively, the lack of bottom coverage improvement with bias voltage may have been due to increased resputtering effects. The increase in bottom coverage of Cu with increasing pressure was also found by other researchers.[27] They also estimated ionization ratio from a computer simulation, and found that it correlated well with observed bottom coverage (Fig. 10).[27] The data in the figure were obtained from 5:1 aspect ratio vias, with no applied bias to the substrate.

If metal ions arrive perfectly normal to the substrate, and in the absence of resputtering effects, the bottom coverage will be 100% and independent of aspect ratio of the feature. In reality, the metal ions will always have some

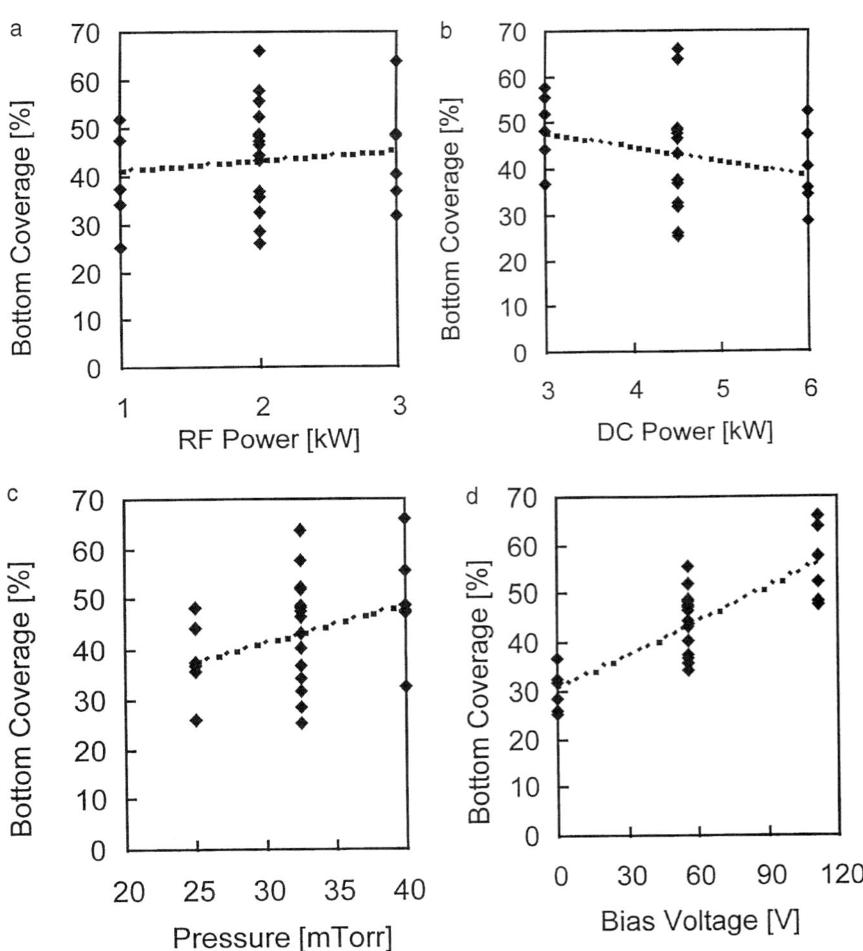

FIG. 7. Bottom coverage of Ti as a function of (a) pressure, (b) RF coil power, (c) DC target power, and (d) wafer bias obtained in an I-PVD source with an internal coil. The via size is $0.35 \times 1.2\,\mu m$ (from Forster *et al.*[13]).

off-normal velocity component. This will make bottom coverage a function of aspect ratio, as shown in Fig. 11,[24] generated for deposition of Ti into vias. This figure also shows how the bottom coverage obtainable with I-PVD compares with the bottom coverage obtainable with other technologies. Curves for deposition of Ta, TaN, and Cu by I-PVD into trenches show a similar behavior.[28] Another curve of bottom coverage as a function of aspect ratio, for Ti deposited into vias, was reported by Yoo *et al.*[29] The

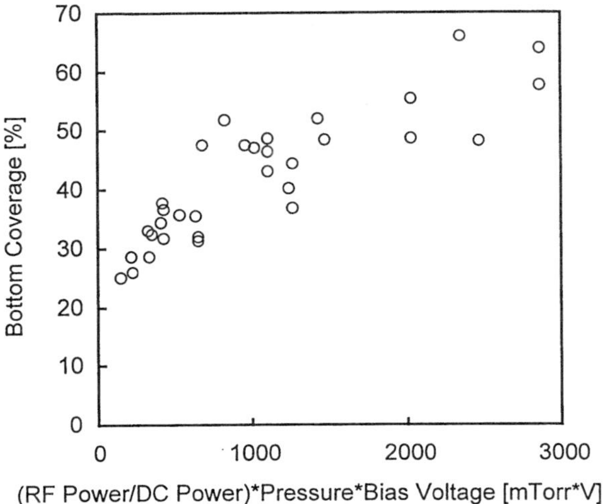

FIG. 8. Bottom coverage of Ti as a function of RF power/DC power × pressure × bias. Data points are the same as those in Fig. 6 (from Forster *et al.*[13]).

data show decreasing bottom coverage with increasing aspect ratio. The bottom coverage with applied wafer bias is higher than the bottom coverage without bias, which is in turn higher than the bottom coverage achievable with collimated Ti. The increase in bottom coverage with applied wafer bias became greater with increasing aspect ratio. This demonstrates the narrowing of the ion angular distribution function at the wafer with application of bias. This same effect was reported by Lo *et al.*[30] There is also an improvement in bottom coverage when using 13.56-MHz bias frequency compared to a 400-kHz bias frequency. Effects of transit time of ions through the sheath can cause a narrowing of the ion energy distribution, and hence the ion angular distribution, as the bias frequency is increased.[31]

2. *Sidewall Coverage, Overhang, and Faceting*

Sidewall coverage, overhang, and faceting are all influenced by resputtering of the deposited film by incoming ions. Reduction of SEM data to obtain quantitative sidewall coverage, overhang, and faceting data is not easy because the sidewall thickness is usually very thin, and there does not exist a universally accepted quantitative definition of either overhang or faceting. Thus, there does not exist a large amount of quantitative data.

Sidewall coverage is also determined by the directionality of the incoming metal. Metal that arrives close to normal can reach the contact sidewall.

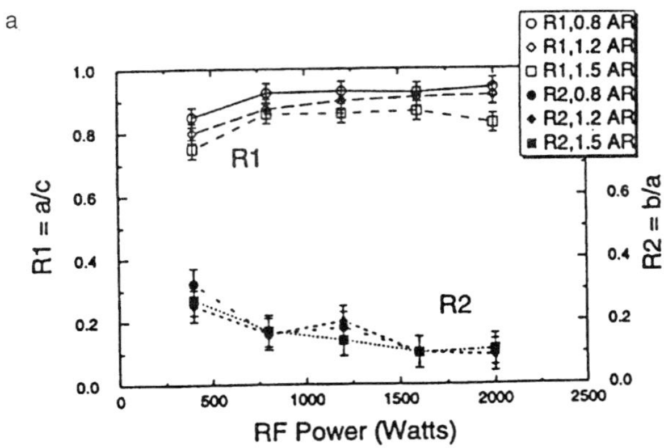

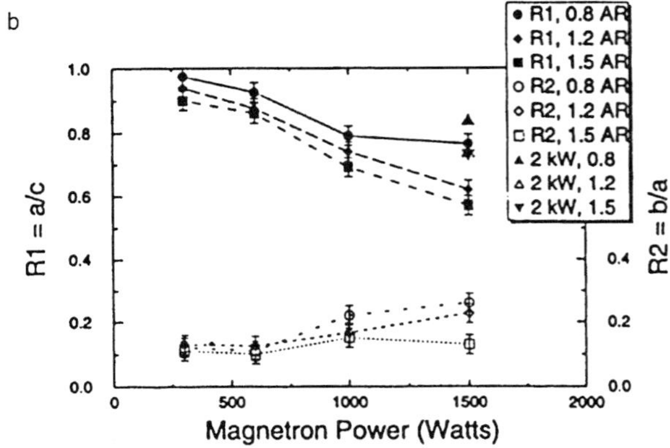

FIG. 9. Bottom coverage and sidewall coverage of Cu, for several trench aspect ratios, as a function of (a) RF power and (b) DC target power for an I-PVD source with internal coil (from Nichols *et al.*[26]).

However, the sidewall deposition due to this off-normal flux decreases toward the bottom of the contact. The sidewall coverage can be enhanced by the resputtering of material in the contact. This resputtered material will most likely arrive at a contact sidewall and redeposit there. This sidewall deposition is greatest at the bottom of the contact and decreases toward the top of the contact. The combination of off-normal flux and resputtered flux will give rise to sidewall coverage that is at a minimum in the middle of the contact and increases toward the bottom and top of the contact, as is shown in Fig. 5. In the absence of resputtering and for a unity sticking coefficient,

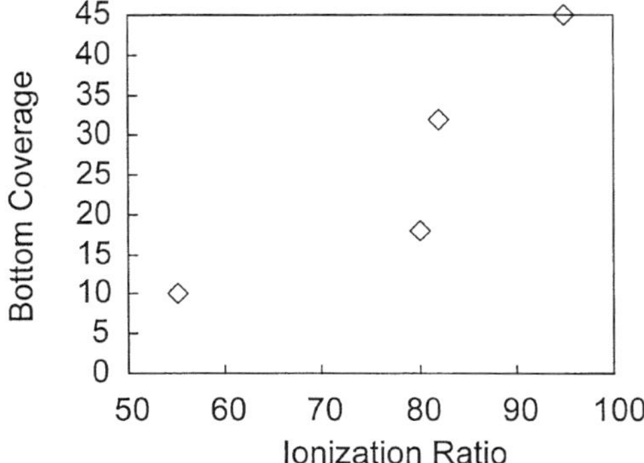

FIG. 10. Bottom coverage of Cu in an I-PVD source as a function of estimated ionization ratio. The via aspect ratio is 5:1 (from Zhang *et al.*[27]).

any increase in sidewall coefficient comes at the expense of bottom coverage, as demonstrated in Fig. 6. This is also borne out experimentally; the data in Figs. 9a and 9b show the bottom coverage and sidewall coverage varying inversely with each other. Bottom coverage increases with increasing RF coil power and decreasing DC target power, whereas sidewall coverage decreases with increasing RF coil power and decreasing DC target power. The data in Fig. 9 do not differ greatly from the theoretical curves in Fig. 6b, indicating that little resputtering occurred during these experiments. As

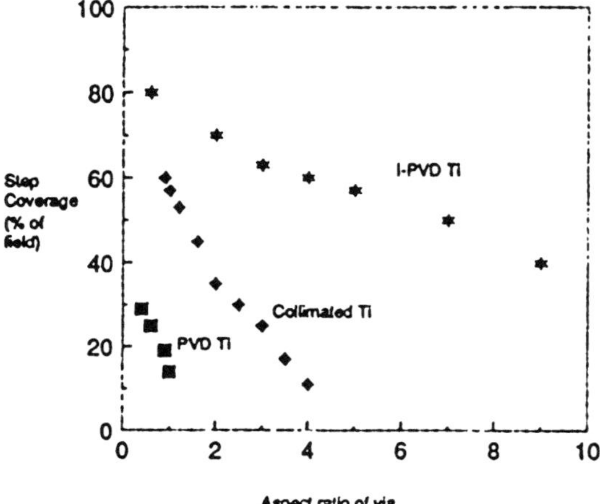

FIG. 11. The bottom coverage of Ti deposited by I-PVD, collimated PVD, and standard PVD as a function of aspect ratio (from Rossnagel[24]).

aspect ratio increases, sidewall coverage becomes more difficult to attain. In Fig. 11, the bottom coverage in a 7:1 aspect ratio feature is ~50%. If there were no resputtering, the sidewall coverage would be limited to ~2%. Therefore, resputtering is essential to achieving adequate sidewall coverage. For example, a dramatic increase in sidewall coverage in 1.6:1 aspect ratio contacts was observed as bias power was increased from 0 W (no applied bias) to 150 W.[32] The application of bias to I-PVD sputtering of Ta increased sidewall coverage at the bottom of a 0.25-μm-wide, 5:1 aspect ratio trench by almost a factor of two, at the expense of bottom coverage.[33] The sidewall coverage in a trench of Ta, TaN, and Cu deposited by I-PVD was reported by Moussavi *et al.*[28] as a function of trench width. As expected, sidewall coverage decreased with decreasing trench width. Unfortunately, the authors did not include process conditions, making it impossible to estimate the effects of resputtering.

It is possible to use a profile simulator to predict the effects of resputtering on sidewall coverage. Successful predictions require knowledge of ion flux, sputter yield as a function of ion energy and incidence angle, angular distribution of sputtered material, and sticking probability. Until recently, knowledge of most of these parameters was very limited. New molecular dynamics simulations are providing the data required for successful profile simulation.[34,35] Figure 12 shows how the sidewall thickness obtained from a simulation increases with increasing sputter yield of the deposited film.[22]

Overhang is due to deposition of off-normal metal flux at the very top corner of the contact. It is undesirable because it hinders filling of the contact. The most effective method for reducing overhang is to bevel the corners of the contact prior to metal deposition. However, IC designers do not like this solution because it reduces the packing density of the contacts. The overhang can be reduced to some extent by resputtering. It is well-known that corners sputter faster than flat films because the sputter yield of most materials as a function of ion incidence angle has a maximum between 45 and 60°. However, too much resputtering can actually increase overhang because the resputtered material tends to land on the opposite sidewall near the top of the contact (see Fig. 5). The maximum in sputter yield as a function of angle also leads to faceting (Fig. 5), where the top corner of the contact has a bevel or facet. The effect of resputtering on faceting and overhang has been studied by computer simulation of profile evolution[22] (see Fig. 12). The length of the beveled face increases with increasing sputter yield of the deposited material. The overhang is particularly important for fill application because any overhang will tend to cause premature closing of the feature and result in a void.

Experimental studies using I-PVD to fill features with Al or Cu have shown how overhang and faceting increase with increasing ion bombard-

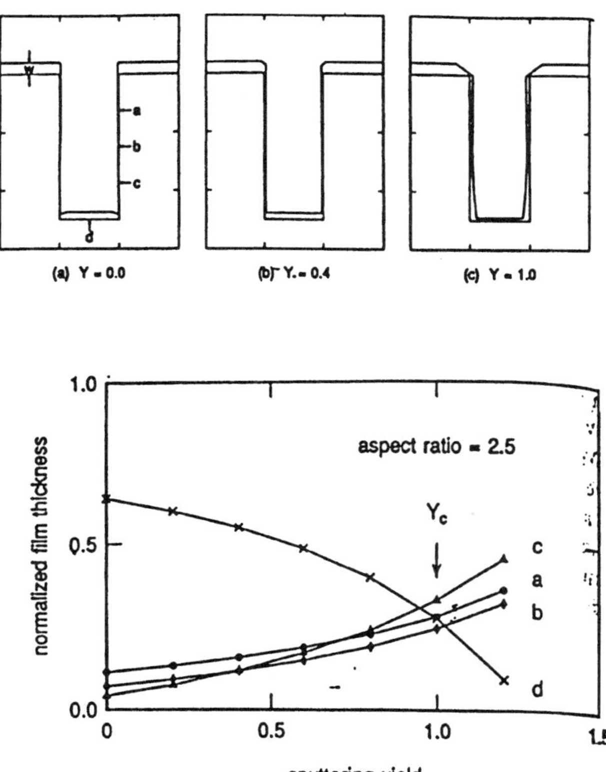

FIG. 12. Simulation showing effect of sputter yield on sidewall coverage (from Hamaguchi and Rossnagel[22]).

ment,[23,36] consistent with theoretical results from profile evolution simulations.[22] Figure 13a[23] dramatically shows how faceting increases with increasing bias voltage (ion energy), with all other plasma parameters held constant. Faceting also increases when the RF coil power (ion flux) increases (Fig. 13b[23]), with all other plasma parameters including wafer bias voltage held constant. An attempt to correlate faceting with process parameters during reactive sputtering of TiN is shown in Fig. 14. In this experiment, target power and bias voltage were held constant (at 5 kW and 100 V, respectively). Pressure, RF coil power, and the Ar to N_2 ratio were varied. It was found that the degree of faceting was inversely related to target voltage. In Fig. 14, the degree of faceting is quantitatively defined as the ratio between the length along the facet and the film thickness. Under constant target power, the target voltage is inversely related to the plasma

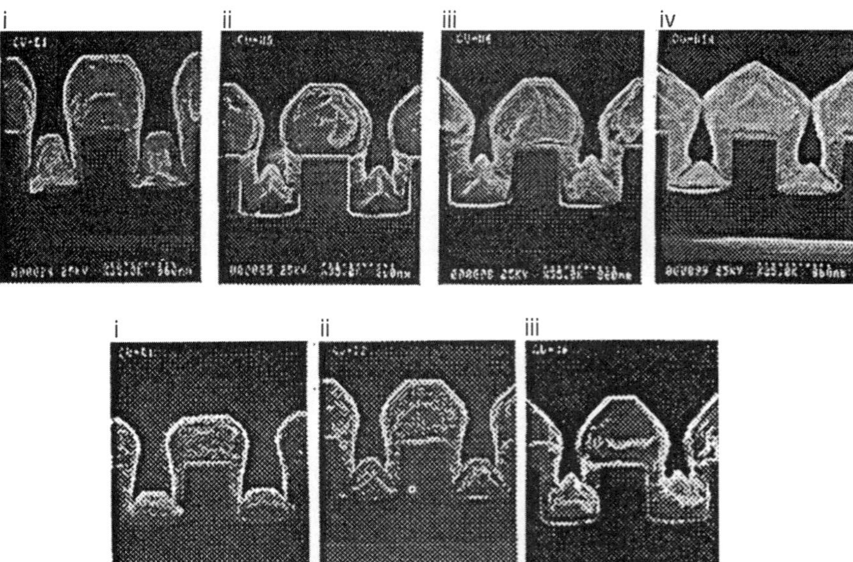

FIG. 13. Faceting of Cu deposited by I-PVD as a function of (a) bias voltage and (b) RF power. In (a), RF power is 1000 W and bias voltage is (i) −5 V, (ii) −20 V, (iii) −30 V, and (iv) −50 V. In (b), bias voltage is −30 V and RF power is (i) 300 W, (ii) 600 W, and (iii) 1000 W (from Cheng *et al.*[23]).

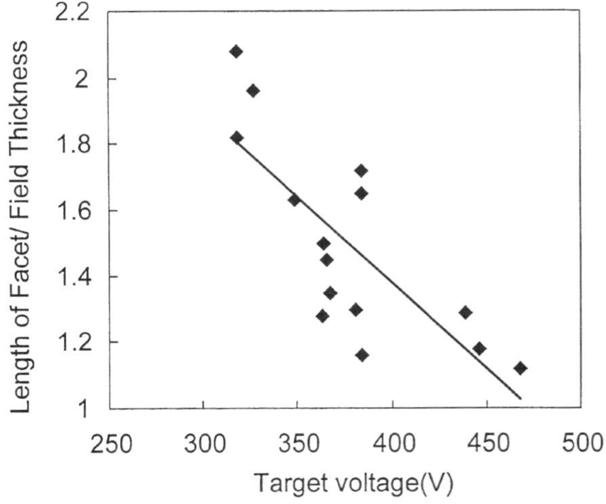

FIG. 14. Faceting of TiN deposited by I-PVD as a function of target voltage. Degree of faceting is defined as the ratio between the length of the facet face and the film thickness on the field.

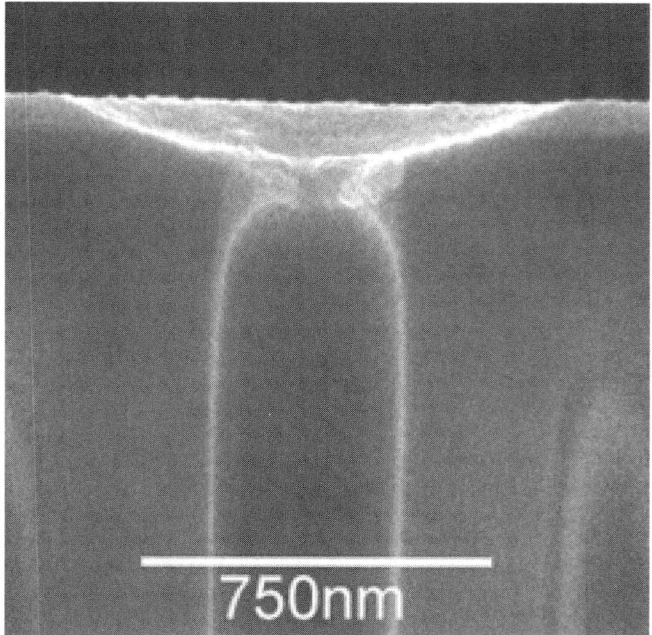

FIG. 15. Example of deposition of Ti with an I-PVD source at low deposition rate and high ion bombardment flux. Note the severe faceting.

density. Thus, the data indicate that if the bias voltage is held constant, faceting increases with increasing plasma density, which is consistent with the data from Ref. 23. Figure 15 shows a SEM of a wafer purposely processed under high bias and high plasma density conditions in order to exaggerate the formation of overhang and faceting. Under these extreme conditions, the facet is very pronounced, and the feature is almost completely closed off.

C. FILM PROPERTIES

This chapter has emphasized the ability of I-PVD sources to fill high aspect ratio features. However, there is another possible advantage of using an I-PVD source to deposit metals. Since the deposition rate, ion-to-neutral ratio, and ion bombardment energy can all be independently controlled, it should be possible to control film properties to a degree not possible with standard PVD sources. In a standard PVD source, it is very difficult to control film microstucture. It has been shown that deposition temperature influences microstructure,[37] as does bombardment by energetic particles.[38] Ion flux can be controlled to some extent by magnetic field geometry,[39] and ion bombardment energy can be controlled via wafer bias. However, the magnetic field geometry is constrained by the need to deposit uniformly over the wafer surface. This constraint greatly limits the ability to control ion bombardment. An I-PVD source offers control over bombardment energy (through the wafer bias), ion flux (through RF power), and ion-to-neutral ratio (through a combination of RF coil power, pressure, and DC target power).

The film composition of reactively sputtered compound films will vary according to process variables such as reactive gas content in the discharge and perhaps ion bombardment. The following review will begin with an overview of film composition in reactively sputtered films and then present results on crystal orientation, grain size, stress, film roughness, and film resistivity for a variety of materials.

1. Film Composition

Film composition is only important for compound films. If no contaminants exist in the source or the sputter target, films such as Ti, Cu, and Al will exhibit no film composition variations. Some films used in microelectronics manufacturing, such as TiN and TaN, consist of a compound. Deposition of such films generally occurs with reactive sputtering. A target of the pure metal constituent is sputtered in an ambient consisting of a mixture of a noble gas (usually Ar) and the reactive component (usually N_2). The addition of a reactive gas makes understanding of the discharge more complex. A brief summary of the operation of a reactive sputter source will be given before discussing film composition of reactively deposited films.

If the sputter yield of the pure metal differs from that of the metal nitride, a phenomenon referred to as "hysteresis" occurs in the graph of pressure vs N_2 gas flow[40] (Fig. 16). With N_2 gas flow set to zero, the pressure is determined solely by the Ar flow rate and Ar pump speed. Initially, as N_2 flow is increased, the reaction between fresh deposited metal and the added N_2 will consume all the N_2 before it can reach the target. Due to the almost complete gettering of the nitrogen by the deposited metal, the pressure does not rise appreciably as N_2 flow increases. In this regime, the source is said to be operating in the "metallic" mode. Eventually, as the N_2 flow is increased further, more N_2 enters the discharge than the deposited metal can consume. Nitrogen can now reach the target in which it reacts with the target surface. As the metal nitride compound forms on the target surface, the sputter rate drops, reducing the amount of deposited metal that can

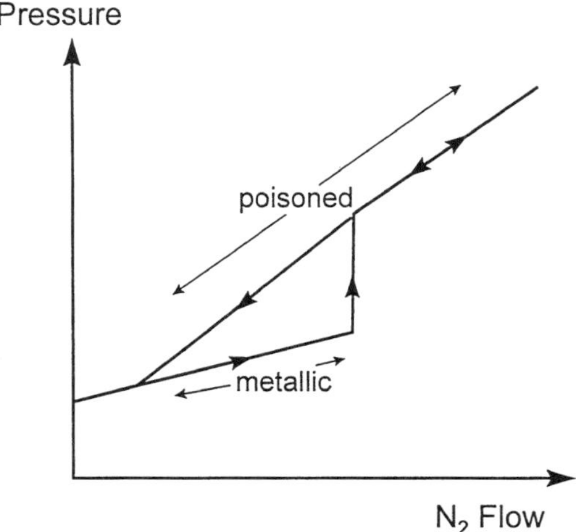

FIG. 16. Schematic hysteresis curve of chamber pressure vs N_2 flow for reactive sputtering.

react with nitrogen. Now there is an excess of nitrogen and the pressure increases drastically. Additional increases in N_2 flow cause an increase in pressure in accord with the flow and the pump speed. In this regime, the source is said to be operating in "poisoned" mode. The actual layer of metal nitride on the target is very thin (~50 Å), and it is not visible to the naked eye. If N_2 flow is reduced, the pressure decreases in accord with the flow and pump speed, and the target remains poisoned, even if the flow is reduced below the initial transition point between metallic and poisoned mode. Eventually, if the N_2 flow is reduced sufficiently, the target will denitride. Hysteresis can also be observed in the plot of target voltage vs N_2 flow, if the metal has a secondary electron emission coefficient different from that of the metal nitride. Titanium nitride exhibits the hysteresis effect because TiN has a sputter yield ranging from one fifth to one third of Ti. Interestingly, TaN does not exhibit hysteresis; TaN has approximately the same sputter yield as Ta.

The film composition can change with N_2 flow in a reactive deposition process. It has been shown that the resistivity of TiN deposited by an I-PVD source operating in the poisoned mode increases with increasing N_2 flow.[41] Interestingly, the film composition varies very little as the source switches from metallic to poisoned mode. The ratio between Ti and N in the film remains near 1:1 within 5%.

The composition, film resistivity, and the lattice structure of the TaN film depend strongly on the nitrogen content of the discharge.[42,43.] As nitrogen is added to a Ta sputtered by an Ar discharge, the deposited material changes from β-phase Ta (N_2/Ar ratio = 0) to bcc Ta (N_2/Ar ~ 0.1) to amorphous or nanocrystalline Ta_2N (N_2/Ar ~ 0.5) to amorphous or nanocrystalline TaN (N_2/Ar > 0.8).[42,43] The resistivity of the TaN_x deposited by I-PVD changes from ~150 $\mu\Omega$-cm (β-phase Ta) to ~60 $\mu\Omega$-cm (bcc-Ta) to ~200 $\mu\Omega$-cm (either nanocrystalline or amorphous Ta_2N or TaN).[42,43]

2. *Crystal Orientation*

The crystal orientation of a film is measured by X-ray diffraction. According to Ying *et al.*[44] (p. 3007), "Nearly all vapor phase deposited metallic films exhibit a strong out-of-plane crystallographic orientation. The texture strongly influences film performance. Body centered cubic films typically exhibit (110) out-of-plane orientations, while face centered cubic films commonly have (111) textures." Performance of metal films for microelectronics applications is important because the resistance of Al interconnects to electromigration has been found to depend strongly on the crystal orientation of the Al.[45,46] The texture of the Al, in turn, has been found to depend strongly on the texture of the underlying layers, usually Ti and/or TiN.[47]

The crystal orientation of Ti films deposited by sputter deposition is usually either (011) or (002). It has been shown that the orientation depends on the level of trace contaminants.[48] It has also been reported that orientation with standard PVD is primarily (002), changing to more (011) if a collimator is used.[49] However, there is also a conflicting report that claims Ti crystal orientation is (002), regardless of whether a collimator is used or not.[50]

The crystal orientation of Ti films deposited by an I-PVD source is primarily (002).[29,51–53]. If high bias is used during deposition, it is possible for the film to appear more randomly oriented, as shown in Fig. 17.[53] Perhaps excessive ion bombardment prevents formation of grains with distinct orientation.

The crystal orientation of TiN films deposited by sputter deposition is usually (111), but other orientations have been observed.[54] The orientation depends strongly on deposition conditions, especially ion bombardment energy and ion flux.[39] The crystal orientation also depend on the substrate underlying the film. The orientation of TiN deposited by I-PVD onto (002)-oriented Ti films was found to be primarily (111).[55]

The crystal orientation of TiN films deposited by an I-PVD source also depend strongly on ion bombardment at the substrate, as shown in Figs. 18a–18d.[56] If deposition is performed on oxide wafers, then at low RF coil power and no applied bias (low ion flux, ion energy ~30 V), the preferred orientation is (111) (Fig. 18a). As RF coil power is increased, with no

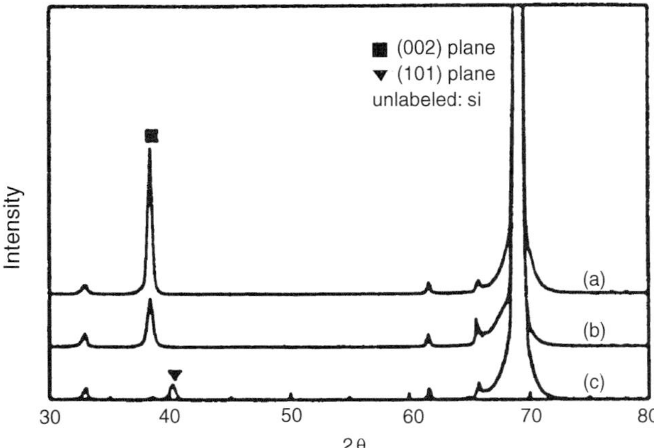

FIG. 17. X-ray diffraction (XRD) curve of Ti deposited by I-PVD; (a) no bias; (b) 150W bias; (c) 300W bias (from Lee *et al.*[53]).

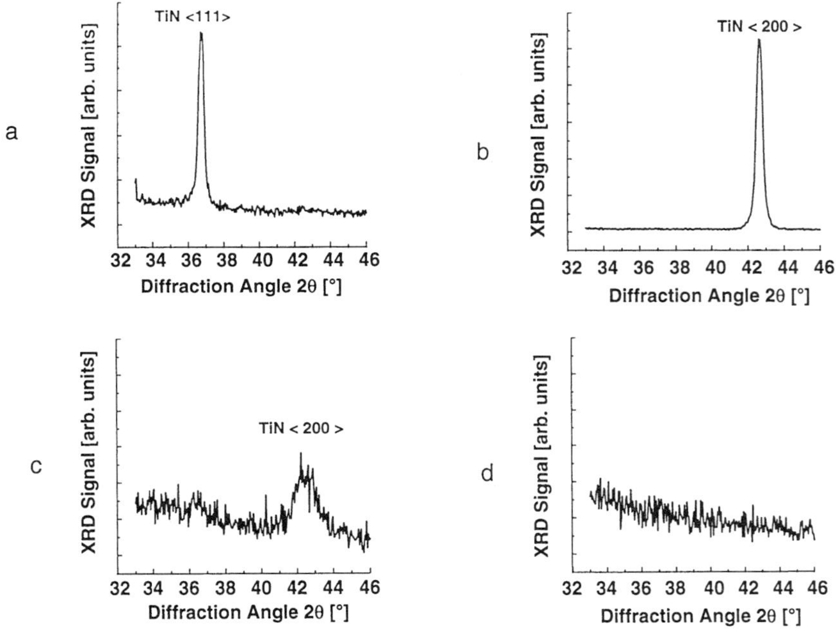

FIG. 18. X-ray diffraction (XRD) curve of TiN deposited by I-PVD, as a function of ion bombardment; (a) 1-kW RF coil power, ion energy ~30 V; (b) 2.5-kW RF coil power, ion energy ~30 V; (c) 2.5-kW RF coil power, ion energy ~90 V; (d) 2.5-kW RF coil power, ~150 V ion energy (from Tanaka *et al.*[56]).

applied bias (high ion flux, ion energy ~30 V), the orientation changes to (200) (Fig. 18b). At high RF coil power, increasing bias to −60 V (ion energy ~90 V) leads to reduction in X-ray diffraction (XRD) peak intensity (Fig. 18c), indicating a loss in crystalline structure due to ion bombardment. If bias is increased to −120 V (ion energy ~150 V), no discernible peaks are visible in the XRD spectrum (Fig. 18d), indicating that the film appears amorphous. The change from crystalline structure to an apparent amorphous structure of TiN deposited by I-PVD with application of wafer bias has been confirmed by other researchers.[29]

The orientation of Al or AlCu films is primarily (111). For best resistance to electromigration, the width of the (111) XRD peak should be narrow.[45,46] The full-width, half-maximum (FWHM) of the XRD (111) peak of Al deposited by an I-PVD source has been shown to be narrower than that for Al deposited by a collimated sputter source.[57] Data from an ionized Al deposition source, consisting of an Al evaporator operating into an arc plasma source, show that the FWHM of the XRD (111) peak decreases as ion energy is increased.[58] It has been proposed that the increased energy available at the wafer surface due to ion bombardment aids in the growth of (111) oriented grains.[58] As mentioned earlier, the texture of Al thin films also depend strongly on the texture of the underlying material. The data in the previous two examples were obtained by depositing Al onto a wafer with an oxide layer on the surface. In an actual IC, the Al would be deposited onto a Ti or TiN layer. It has been shown that the FWHM of the Al (111) can be reduced if the underlying Ti layer is primarily (002).[47] Figure 17 has shown that Ti deposited with an I-PVD source is primarily (002) oriented. The FWHMs of Al deposited on Ti films deposited by an I-PVD source have been shown to be smaller than the FWHMs of Al films deposited on Ti films deposited by collimated sputtering.[59]

3. *Grain Size and Film Roughness*

The grain size of a thin film is usually measured by examining either SEM or TEM images of the film. The film roughness is usually measured by atomic force microscopy (AFM). The grain size and roughness of a film can depend on the deposition conditions, as texture can be influenced by resputtering of the deposited material and by self-shadowing of faster growing grains.[44]

The effect of ion flux and ion bombardment energy on the grain size of TiN films deposited by an I-PVD source is shown in Figs. 19a–19d.[56] The process conditions of the various images correspond to the process conditions in Figs. 18a–18d. The figure shows AFM images looking top down onto the film surface (Fig. 19a) and TEM cross sections of the film (Fig.

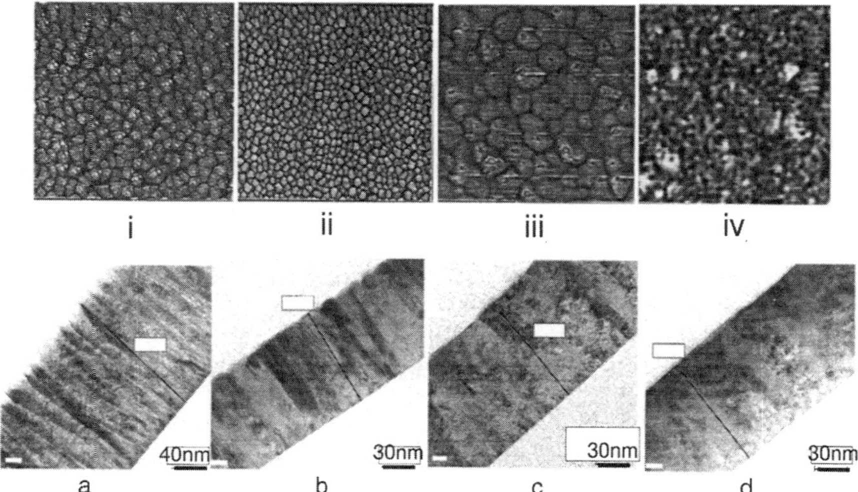

FIG. 19. (a) Surface roughness (AFM image: area is 500 × 500 nm) and (b) TEM cross section of TiN films deposited by I-PVD. Conditions are the same as those for Fig. 18; (i) 1-kW RF coil power, ion energy ~30 V; (iii) 2.5-kW RF coil power, ion energy ~30 V; (iii) 2.5-kW RF coil power, ion energy ~90 V; (iv) 2.5-kW RF coil power, ~150 V ion energy (from Tanaka *et al.*[56]).

19b). At low ion flux and low ion bombardment energy, the TiN film exhibits needle-like grains with porous features. The grain size is ~200–300 Å. As ion flux increases, but ion bombardment energy is kept low (~30 V), the structure becomes columnar with distinct grains. The grain size remains 200–300 Å. The TiN grain size increases to ~700 Å if ion bombardment energy is increased to ~90 V. Grain structure is difficult or impossible to discern if the ion bombardment energy is increased further (>150 V). This is consistent with the loss of XRD signal shown in Fig. 18.

Wafer bias also has a similar effect on Ti films deposited by I-PVD sources. Increasing wafer bias leads to an increase in film roughness[51,53] and to less pronounced grain features.[53]

4. *Resistivity and Stress*

The microstructure of a thin film will affect macroscopic film parameters. Film resistivity and stress are of great concern in microelectronic applications. The resistivity of a liner or barrier layer will affect the resistance of the via or contact. Resistivity of TiN is determined by film structure and film stoichiometry. Resistivity of Ti and Al is mainly determined by film structure. Resistivity of a thin film can also be affected by incorporation of "foreign" atoms. The film stress can impact device reliability. If stress is too high, then peeling or cracking may occur. Stress can also affect the electromigration resistance of Al lines. Due to generation of interstitials by "ion peening," it is expected that stress would become more compressive as ion bombardment increases.[38] The stress will also depend on wafer temperature during deposition.[38]

The wafer bias has been shown to have the most dramatic effect on stress in Ti films.[51,52] Strangely, the data indicate that stress in Ti films becomes more tensile with increasing wafer bias.[51] This result may be due to wafer heating effects not accounted for in the experiment.

The stress of TiN films deposited by I-PVD was found to become more compressive with increasing wafer bias,[54,56] consistent with data from non-I-PVD sources.[60] The resistivity of TiN deposited by I-PVD is shown in Fig. 20.[56] The resistivity follows the trends of film microstructure. As the film becomes denser and grain size increases, resistivity decreases. At high ion bombardment energies, resistivity increases. This may be due to the loss of grain structure or perhaps the measured increase in Ar incorporation at high ion bombardment energy. The resistivity of TiN deposited by I-PVD also depends on the N_2 flow during processing.[41] In general, increasing the N_2 flow increases resistivity. This result is not surprising since a similar dependence is observed for standard PVD-deposited TiN.

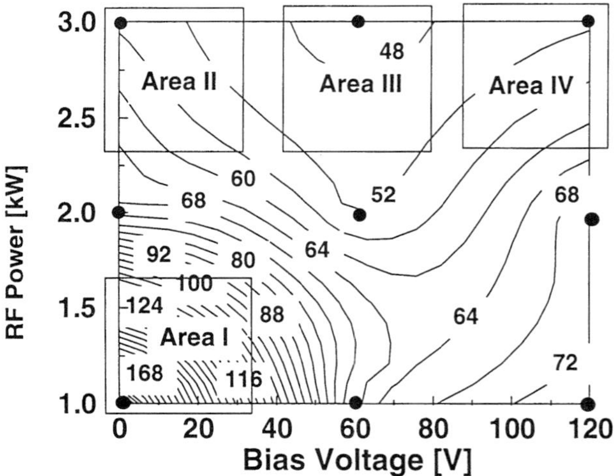

FIG. 20. Resistivity of TiN film as a function of applied bias voltage and RF coil power. Note that plasma potential is ~30 V so that ion bombardment energy is approximate applied bias voltage +30 V (from Tanaka *et al.*[56]).

The stress of AlCu films deposited by I-PVD was found to be slightly more tensile than that of films deposited by collimated sputtering.[55] The authors surmised this to be due to wafer heating during deposition.

IV. Applications

The I-PVD source has been used for several applications in microelectronic manufacturing. Process qualification for manufacturing involves several steps. The new process must be integrated with all prior and subsequent process operations. For example, introducing an I-PVD liner/barrier process requires that subsequent plug fill operations can provide mechanically stable fills. In the case of W plug fill, the W must be shown to adhere well to the I-PVD liner/barrier with no peeling or attack of the liner, and the W fill must show no keyholes. In the case of Al plug fill, the fill must show no voids, and at contact level there can be no penetration of the Al through the I-PVD barrier into the Si devices. Finally, introduction of a new process requires that electrically testable structures be built with the process. The simplest back ends of line structures are contacts or vias, and electrical testing consists of measuring resistance of the contacts or vias. At contact level, leakage current or the contacts can also be measured.

A. LINERS AND BARRIERS FOR W PLUG INTERCONNECTS

Liners and barriers for W plug interconnects consist of a thin layer of TiN deposited over a thin layer of Ti. The initial Ti layer is used as an adhesion promoter, and the TiN is a barrier that protects the Ti from the corrosive WF_6 used during W plug fill. A continuous layer of TiN is required for good barrier performance. Failure of the barrier leads to peeling of the TiN layer, illustrated by the SEM in Fig. 21a.[61] In general, a thinner barrier/liner is more prone to failure but is more likely to allow good fill of W. Therefore, deposition methods which offer high bottom and sidewall coverage are desirable.

Successful integration of an I-PVD Ti/TiN liner/barrier with a CVD-W plug fill requires optimization of both the I-PVD and the CVD–W process[61] (a representative TEM of a successful fill is shown in Fig 21b). Inclusion of an anneal step between the TiN and W deposition often improves barrier performance. The CVD–W deposition generally consists of a nucleation step prior to deposition. The fill and barrier integrity can be adjusted by optimizing the nucleation step.[61]

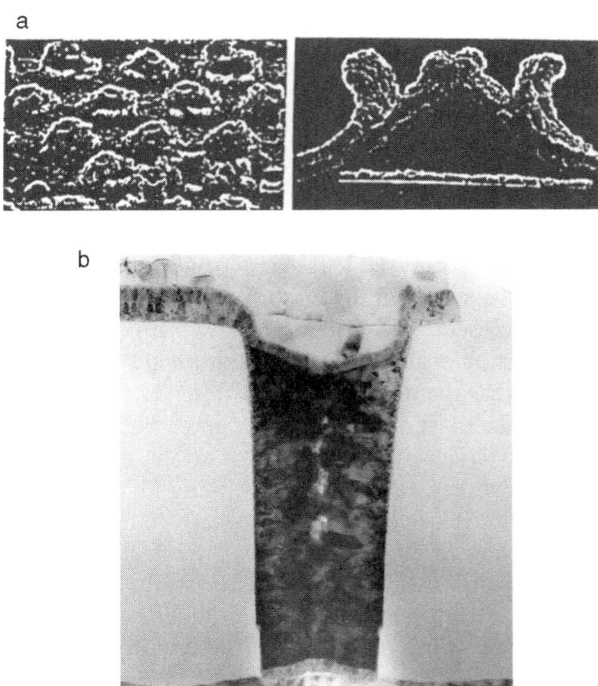

FIG. 21. The importance of proper integration of the TiN and W deposition process. (a) SEM image of a TiN barrier failure that failed during tungsten CVD deposition (from Tanaka *et al.*[61]). (b) TEM of a successful TiN and W integration (from Dixit *et al.*[62]).

As mentioned previously, the resistance of a silicided contact will depend on the amount of Ti deposited into the contact prior to thermal annealing. This is confirmed by reviewing electrical data from contacts receiving a Ti liner, a TiN barrier, and CVD–W fill. A comparison between I-PVD-deposited Ti and collimated PVD-deposited Ti, in 3:1 aspect ratio contacts using a CVD–TiN barrier (Fig. 22),[62] shows the median resistance obtained with 200 Å I-PVD Ti to be comparable to the resistance obtained with 600 Å collimated Ti. This indicates that the I-PVD bottom coverage is approximately three times the collimated Ti bottom coverage. In addition, the cumulative distribution of the I-PVD is steeper than the distribution of the collimated Ti, indicating less variation in contact resistance across the wafer. Another study showed that 200-Å thick Ti deposited by I-PVD resulted in a significantly lower contact resistance in 0.4-μm, $>$5:1 aspect ratio contacts than did 600-Å-thick collimated Ti.[29] Application of bias to

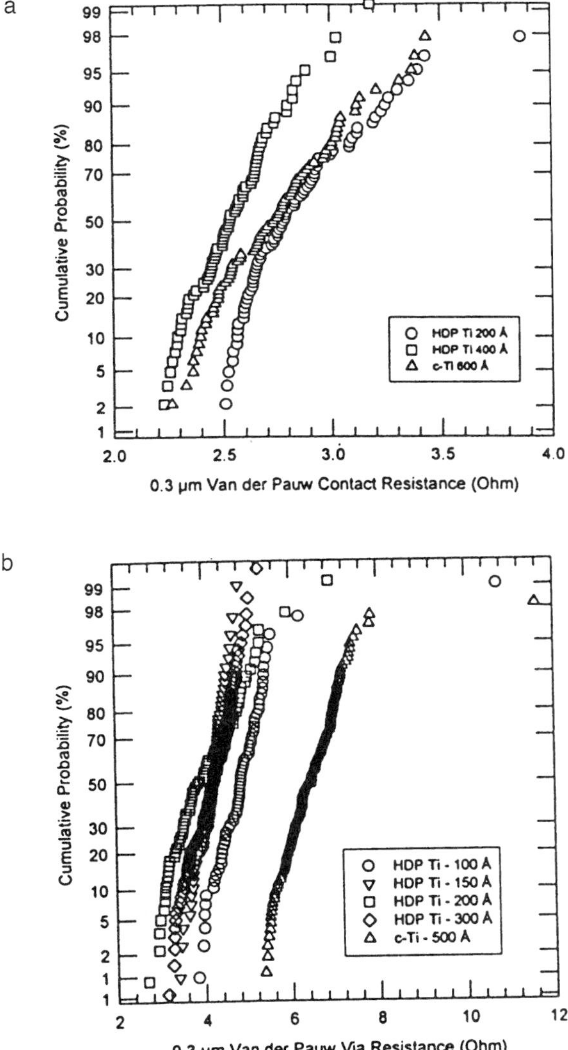

FIG. 22. Cumulative distributions of contact (a) and via (b) resistance of holes with I-PVD Ti liner, CVD–TiN barrier, and CVD–W plug (from Dixit *et al.*[62]).

the wafer during deposition of Ti by I-PVD resulted in an improvement in contact resistance of ~20–30% compared to Ti deposited without bias (Fig. 23).[29] A similar improvement in contact resistance when using I-PVD-deposited Ti as opposed to collimated Ti was reported for 5:1 aspect ratio contacts.[51] The data from these studies clearly indicate that contact resis-

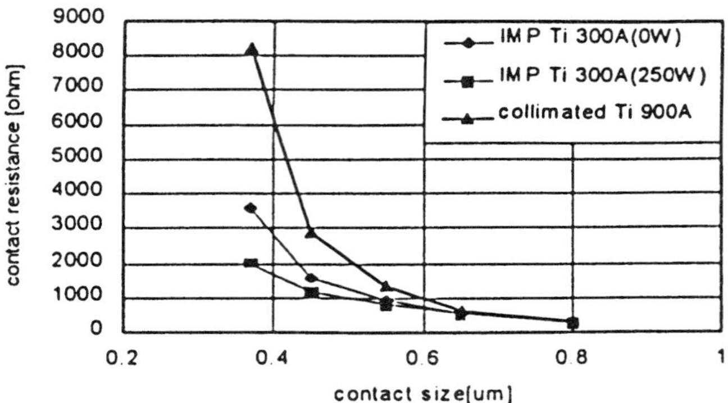

FIG. 23. Contact resistance as a function of contact diameter, comparing 900 Å of collimated PVD-deposited Ti with 300 Å of I-PVD-deposited Ti. The curve with "(0 W)" has no applied wafer bias; the curve with "(250 W)" has 250-W applied wafer bias (from Yoo *et al.*[29]).

tance is directly related to the amount of Ti deposited at the bottom of the contact structure. Therefore, the improvement in bottom coverage possible with I-PVD over standard or collimated PVD can lead to lower contact resistance for high aspect ratio contacts. This also implies that for a given contact resistance, use of I-PVD will require a thinner Ti deposition than use of collimated or conventional PVD, thus easing integration requirements with the subsequent W fill.

The higher bottom coverage exhibited by I-PVD sources allows the field thickness of the liner and barrier layers to be thinner. This eases the requirements of process integration with the CVD–W deposition, especially during W etchback. The effect of the TiN barrier on electrical performance is less clear. A thinner barrier should lead to lower resistance. However, the barrier performance is to first order more dependent on barrier thickness than on deposition method. Nevertheless, TiN barriers deposited by I-PVD show lower contact or via resistance than those deposited by collimated or conventional PVD.[29,62,63] Perhaps the resistance is also determined by the amount of overhang of the barrier.

A reduction in contact resistance when using 300-Å I-PVD-deposited TiN rather than 900-Å collimated PVD-deposited TiN in 0.4-μm, >5:1 aspect ratio contacts was observed by other researchers.[29] The good barrier performance of the thinner I-PVD barrier was attributed to better film properties of the sidewall deposition. A comparison of via resistances between TiN deposited by I-PVD and TiN deposited by standard PVD, as a function of via diameter, is shown in Fig. 24.[64] The via depth is 1 μm. The via resistance when using 200-Å I-PVD TiN is lower than the via resistance

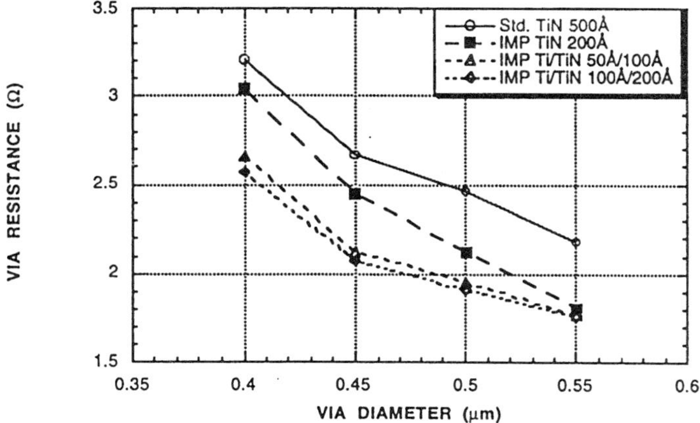

FIG. 24. Via resistance comparison between I-PVD and standard PVD glue layers. The vias were filled by CVD–W (from Bothra *et al.*[64]).

when using 500-Å standard PVD TiN. Use of an I-PVD Ti adhesion layer prior to TiN deposition reduces the contact resistance further. Lower contact resistance and lower via resistance for I-PVD-deposited Ti and TiN were observed compared to either standard PVD- or collimated PVD-deposited Ti and TiN.[63] In addition, the variation in contact and via resistance across the wafer for the I-PVD-deposited Ti and TiN was lower than that for either standard PVD- or collimated PVD-deposited Ti and TiN. A comparison of via resistance between I-PVD-deposited TiN and CVD-deposited TiN for W plug application was shown by Wang *et al.*[65] Both TiN layers had an I-PVD-deposited Ti adhesion layer. No significant difference between the TiN deposition methods is evident. The use of a Ti adhesion layer deposited by I-PVD showed less variation in via resistance across the wafer than for a Ti adhesion layer deposited by standard PVD.

B. LINERS AND BARRIERS FOR Al PLUG INTERCONNECTS

Liners and barriers for Al plug interconnects typically consist of Ti and TiN. A barrier is only required at the contact level to prevent Al from diffusing into the silicon devices. The barrier must be able to prevent diffusion of the Al even at the elevated temperature. The difficulty in depositing a reliable diffusion barrier has prevented the use of Al at the contact level for most ULSI applications. Several overviews of diffusion barriers are available.[1,2,66] A Ti liner is used to enhance adhesion of the Al because both metals will react to form a stable product, $TiAl_3$. Very thin layers of Ti are desirable because $TiAl_3$ has a much higher resistivity than Al. Deposition of Al plugs typically involves PVD deposition at elevated temperature to allow the Al to flow into the via. The Ti adhesion layer plays a critical role in the flow properties of the Al. The Ti adhesion layer should coat the entire sidewall of the via continuously to prevent the formation of voids.

There are several methods that use standard PVD to fill small features with Al. The Al can be deposited at elevated temperature (>350°C) to ensure hole filling ("hot Al"). The aluminum can also be deposited at low temperature (<200°C), and then the wafer can be heated to enable hole filling ("reflow Al"). The most advanced method is to deposit a thin seed layer of Al at low temperature (<200°C) and then deposit the remainder of the Al at elevated temperature (<350°C) ("cold–hot Al"). The seed layer enhances adhesion of the Al to the sidewalls of the feature. Murarka[2] and Xu *et al.*[66] discuss the use of hot Al, cold–hot Al, or reflow Al to fill vias in greater detail. Alternatively, it has been proposed that I-PVD could be used to directly fill the contacts or vias.

There are several reasons why deposition of barrier and liner layers by I-PVD for Al plug filling would have an advantage over deposition by traditional means. The enhanced bottom coverage of I-PVD is the most obvious reason. The use of I-PVD might also influence film properties, as mentioned previously, that may improve the flow properties of the Al.

The use of I-PVD to deposit a trilayer Ti/TiN/Ti liner/barrier enabled cold–hot Al plug filling at temperatures lower than those if standard PVD Ti/TiN deposition had been used.[67,68] The ability to fill at lower temperature was attributed to the ability of I-PVD to deposit a smoother TiN film and a reduced overhang due to better step coverage when using I-PVD. The experiment in Ref. 67 deposited Ti/TiN/Ti using a single I-PVD chamber. The uppermost layer of Ti actually contained some nitrogen remaining from the previous TiN deposition and was referred to as TiN_x. Increasing the thickness of the first Ti layer was found to improve the FWHM of the Al (111) rocking curve but did not influence filling. The fill capability improved if the last (uppermost) Ti (TiN_x) layer was thicker and deposited at lower pressure. Figure 25 shows a 5:1 aspect ratio hole filled with cold–hot Al at 425°C after receiving an I-PVD-deposited Ti/TiN/Ti stack.[67]

At the via level, there is no need for a TiN diffusion barrier. Only a Ti wetting layer is needed to enhance adhesion and reflow of Al into the vias. The use of I-PVD-deposited Ti as a wetting layer resulted in a lower via resistance than if collimated PVD-deposited Ti were used.[32] This study also demonstrated the importance of the step coverage of the wetting layer. Use of an I-PVD-deposited wetting layer with applied wafer bias enabled complete Al filling of the vias compared to incomplete filling if no bias was

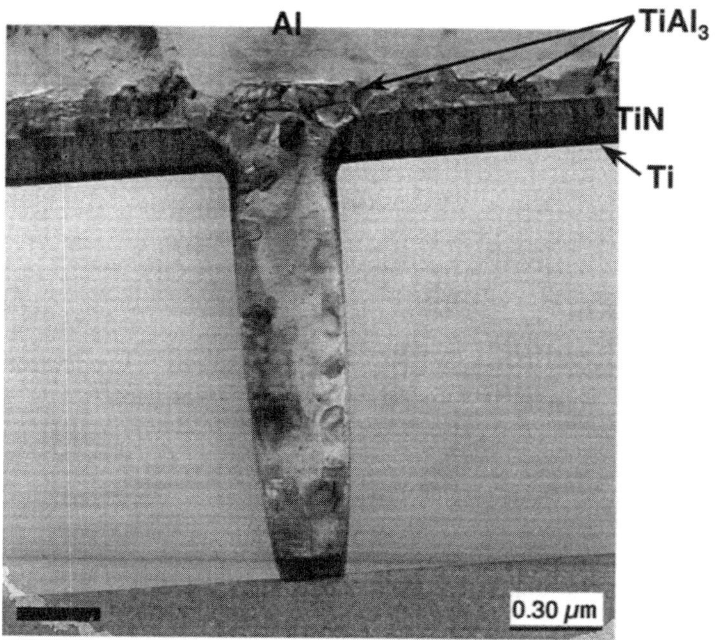

FIG. 25. A 5:1 aspect ratio hole with I-PVD Ti/TiN/TiN_x liner/barrier, filled with "cold-warm" Al (from Chang *et al.*[67]).

applied to the wafer. It was suggested that the enhanced sidewall coverage due to resputtering with applied wafer bias led to better Al fill.

C. LINERS AND BARRIERS FOR Cu INTERCONNECTS

The requirements for liners and barriers for Cu interconnects is more stringent than that for W plug or Al interconnects. Copper can diffuse readily through many materials commonly used in microelectronic manufacturing, including SiO_2. This requires that the Cu be surrounded by a diffusion barrier. Both Ta and TaN have been proposed as Cu diffusion barriers. This section will focus entirely on Ta and TaN diffusion barriers.

The I-PVD deposition of TaN is performed with reactive sputtering. A pure Ta target is sputtered with a plasma consisting of both Ar and N. The film resistivity and the lattice structure of the TaN film depend strongly on the nitrogen content of the discharge.[42,43] As nitrogen is added to a Ta sputtered by an Ar discharge, the deposited material changes from

β-phase Ta (N_2/Ar ratio = 0) to bcc Ta (N_2/Ar ~ 0.1) to amorphous or nanocrystalline Ta_2N (N_2/Ar ~ 0.5) to amorphous or nanocrystalline TaN (N_2/Ar > 0.8).[42,43] The resistivity of the TaN_x deposited by I-PVD changes from ~150 $\mu\Omega$-cm (β-phase Ta) to ~60 $\mu\Omega$-cm (bcc-Ta) to ~200 mΩ-cm (either nanocrystalline or amorphous Ta_2N or TaN).[42,43]

The barrier properties of I-PVD-deposited TaN as a function of film stoichiometry were measured by two methods.[42] For both methods, the film stoichiometry was controlled by varying the nitrogen flow during reactive sputtering of the TaN. An electrical test was used to measure leakage current through a MOS capacitor as a function of time. A mean time to failure was defined as the time for leakage current to increase by two orders of magnitude. One electrode of the capacitor consisted of Cu deposited over a barrier layer. The mean time to fail increased with increasing nitrogen content of the barrier (Fig. 26). Another test involved depositing the barrier layer on a bare Si wafer, followed by a deposition of Cu. After an anneal at 450°C for 30 min, the metal films were removed by wet etching. Another etch was then applied to decorate at Cu that had diffused through the barrier into the Si. Inspection of the decorated wafers by SEM showed that good barrier performance occurred if the TaN_x deposited by I-PVD had $x \gtrsim 0.3$. Another barrier test used secondary ion mass spectroscopy to measure the Cu in the oxide after an anneal showed no detectable Cu level for a TaN barrier but significant Cu levels for a sputtered Ti/TiN barrier.[43]

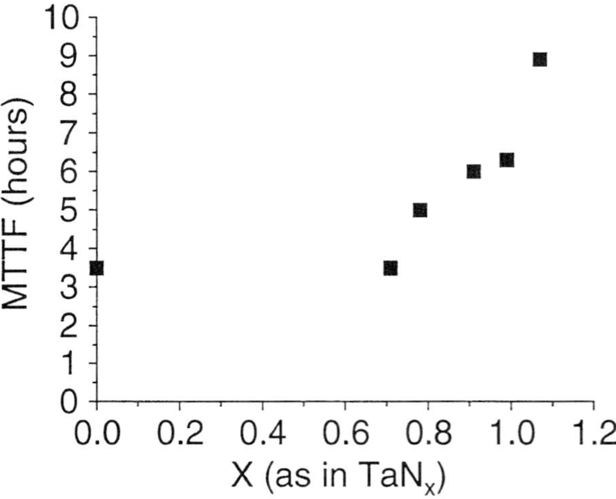

FIG. 26. Leakage current of MOS capacitors as a measure of TaN barrier performance against Cu diffusion (from Sun[42]).

D. SEED LAYERS FOR Cu ELECTROPLATING

Recently, electroplating of copper has been demonstrated as a viable copper fill technique due to its low deposition temperature, high deposition rate, and low cost of consumables.[69] Successful integration of the copper electroplating process into the manufacturing of ICs requires a deposition process for the electroplating seed layer. This seed layer acts as a nucleation layer, and it carries the current that drives the electroplating process. The seed layer needs to be thin, smooth, and continuous.

The sidewall and bottom coverage characteristics of the I-PVD source make it an ideal candidate for Cu seed layer applications. It is necessary to keep the wafer cool ($<100°C$) to prevent the Cu from becoming rough due to "de-wetting."[33,70] Rough and de-wetted films can have discontinuous sidewall coverage, leading to incomplete fill during subsequent electroplating. The high plasma density present in I-PVD sources leads to high ion flux at the wafer, resulting in a high heat flux to the wafer. An ion current density of 20 mA/cm^2, assuming a 30-V sheath voltage at an unbiased wafer, will cause a heat flux of 0.6 W/cm, or a total of ~ 180 W for a 200-mm-diameter wafer. Any additional bias power will heat the wafer further. These high heat fluxes require the use of wafer chucks that can remove the large quantity of heat.

The probability of successful electroplating fill following seed layer deposition increases with increasing sidewall coverage.[33] Features filled by electroplating using an I-PVD Cu seed layer are shown in Fig. 27. The features shown in Fig. 26 also had an I-PVD-deposited TaN barrier layer underneath the Cu seed layer.

Electrical results from vias in low-*k* material filled with I-PVD TaN, I-PVD Cu seed layer, and electroplated Cu have been reported.[71] The resistance and capacitance of the interconnects was lower than those for comparable Al/SiO_2-based interconnects. Low leakage current measurements demonstrated the stability and reliability of the Cu/low-*k* interconnects. Via resistance measurements for vias in SiO_2 filled with an I-PVD-deposited TaN barrier, I-PVD-deposited Cu seed layer, and electroplated Cu fill also showed successful fill.[33]

V. Conclusions

The I-PVD source has progressed from a research tool to a solution for advanced metal deposition in microelectronic manufacturing. In addition to the applications reviewed in this chapter, there are several being pursued

FIG. 27. Copper plug fill using I-PVD Cu seed layer followed by Cu electroplating (from Chin *et al.*[33]).

that may be published by the time this book appears in print. It should be mentioned that the I-PVD source has also been used for deposition of dielectrics.[72]

References

1. S. Wolf, *Silicon Processing for the VLSI Era, Volume 2: Process Integration,* Lattice Press, Sunset Beach CA, 1990.
2. S. P. Murarka, *Metallization: Theory and Practice for VLSI and ULSI,* Butterworth–Heineman, Boston, 1993.
3. A. Denboer, *Semiconductor Int.*, **64** (1994).
4. M. Wittmer, *J. Vac. Sci. Technol. A* **2**, 273 (1984).
5. M.-A. Nicolet, *Thin Solid Films* **52**, 415 (1978).
6. L. Ouellet, Y. Tremblay, G. Gagnon, M. Caron, J. F. Currie, S. C. Gujrathi, and M. Biberger, *J. Vac. Sci. Technol.* B **14** 2627 (1996).
7. R. L. Anderson and J. C. Helmer, U.S. Patent No. 4,995,958 (Feb. 26 1991).
8. T. Smy, K. L. Westra, and M. J. Brett, *IEEE Trans. Electron Dev.* **37**, 591 (1990).
9. T. Heberlein, G. Krautheim, and W. Wuttke, *Vacuum* **42**, 47 (1991).

10. M. J. Grapperhaus, Z. Krivokapic, and M. J. Kushner, *J. Appl. Phys.* **83**, 35 (1998).
11. M. Li. and D. B. Graves, AVS Symposium San Jose, Paper No. ThM6, 1997.
12. J. E. Foster, W. Wang, A. E. Wendt, and J. Booske, *J. Vac. Sci. Technol. B* **16**, 532 (1998).
13. J. Forster, Y. Tanaka, P. Gopalraja, T. Tanimoto, and R. Hofmann, submitted for publication.
14. M. Dickson and J. Hopwood, *J. Vac. Sci. Technol. A* **15**, 2307 (1997).
15. J. Hopwood, *Phys. Fluids* **5**, 1624 (1998).
16. M. Dickson, G. Zhong, and J. Hopwood, *J. Vac. Sci. Technol. B* **16**, 523 (1998).
17. M. Yamashita, *J. Vac. Sci. Technol. A* **7**, 151 (1989).
18. K. Nakamura, Y. Kuwashita, and H. Sugai, *Jpn. J. Appl. Phys.* **34**, L1686 (1995).
19. K. Eng, K. Strohmaier, R. Palmer, B. Stoner, and S. Washburn, *Rev. Sci. Instrum.* **68**, 2381 (1997).
20. M. J. Kushner, personal communication.
21. D. N. Ruzic, personal communication.
22. S. Hamaguchi and S. M. Rossnagel, *J. Vac. Sci. Technol. B* **14**, 2603 (1998).
23. P. F. Cheng, S. M. Rossnagel, and D. N. Ruzic, *J. Vac. Sci. Technol. B* **13**, 203 (1995).
24. S. M. Rossnagel, *J. Vac. Sci. Technol. B* **16**, 2585 (1998).
25. P. L. Ventzek, M. Hartig, V. Arunachalam, D. G. Coronell, and D. Denning, AVS Symposium, Baltimore, Paper No. PS-ThM8, 1998.
26. C. A. Nichols, S. M. Rossnagel, and S. Hamaguchi, *J. Vac. Sci. Technol. B* **14**, 3270 (1996).
27. H.-M. Zhang, I. Hashim, P. Ding, B. Chin, and J. Forster, AVS Symposium, Baltimore, Paper No. PS-ThM9, 1998.
28. M. Moussavi, Y. Gobil, L. Ullmer, L. Perroud, P. Motte, J. Torres, F. Romagna, M. Fayoulle, J. Palleau, and M. Plissonier, IITC Proceedings, p. 295, IEEE, 1998.
29. B. Y. Yoo, Y.-H. Park, H.-D. Lee, J. H. Kim, H.-K. Kang, M. Y. Lee, H. G. Wang, K.-S. Lee, J. VanGogh, C.-H. Chu, S. Lai, B. McClintock, and S. Edelstein, IITC Proceedings, p. 262, IEEE, 1998.
30. S. C. Lo, S. Chiu, K. M. Yin, C. C. Chiang, C. Chen, F. R. Chen, and J. J. Kai, VMIC Proceedings, p. 130, 1998.
31. M. S. Barnes, J. Forster, and J. H. Keller, *IEEE Trans. Plasma Sci.* **PS-19**, 240 (1991).
32. G. Lau, S. Geha, E. Shan, Z. Wu, J. Su, S. Ponnekanti, and G. Yao, VMIC Proceedings, p. 114, 1998.
33. B. Chin, P. Ding, B. Sun, T. Chiang, D. Angelo, I. Hashim, Z. Xu, S. Edelstein, and F. Chen, *Solid State Technol.* (July 1998).
34. A. Kersch and U. P. Hansen, AVS Symposium, Baltimore, Paper No. PS-WeM6, 1998.
35. C. F. Abrams and D. Graves, AVS Symposium, Baltimore, Paper No. PS-ThM11, 1998.
36. Rossnagel, *J. Vac. Sci. Technol. B* **13**, 125 (1995).
37. J. A. Thornton, *J. Vac. Sci. Technol.* **11**, 666 (1974).
38. D. L. Smith, *Thin Film Deposition*, pp. 423–429, McGraw-Hill, New York, 1995.
39. J. Musil and S. Kadlec, *Vacuum* **40**, 435 (1990).
40. D. L. Smith, *Thin Film Deposition*, pp. 479–481, McGraw-Hill, New York, 1995.
41. P. Gopalraja, Y. Tanaka, T. Tanimoto, R. Hofmann, J. Forster, and Z. Xu, paper presented at the MRS Symposium, 1996.
42. B. Sun, Proceedings of the 1997 Advanced Metallization Conference on ULSI Applications, MRS, p. 137.
43. J. Mendonca, R. Venkatraman, G. Hamilton, M. Angyal, B. Rogers, L. Frisa, V. Kaushik, C. Simpson, T. P. Ong, M. Herrick, R. Gregory, T. Remmell, R. Fiordalice, J. Klein, E. Weitzman, T. Chiang, P. Ding, and B. Chin, Proceedings of the 1997 Advanced Metallization Conference ULSI Applications, MRS, p. 741.

44. F. Ying, R. W. Smith, and D. J. Srolovitz, *Appl. Phys. Lett.* **69**, 3007 (1996).
45. S. Vaida and A. K. Sinha, *Thin Solid Films* **75**, 253 (1981).
46. D. B. Knorr and D. P. Tracey, *Appl. Phys. Lett.* **59**, 16 (1991).
47. M. Sekiguchi, K. Sawada, M. Fukumoto, and T. Kouzaki, *J. Vac. Sci. Technol. B* **12**, 2992 (1994).
48. T. Yoshida, S. Hashimoto, Y. Mitsushima, T. Ohwaki, and Y. Taga, *J. Vac. Sci. Technol. B* **16**, 2751 (1998).
49. T. Hara, T. Nomura, R. C. Moseley, H. Suzuki, and K. Sone, *J. Vac. Sci. Technol. A* **12**, 506 (1994).
50. S.-Q. Wang and J. Schlueter, *J. Vac. Sci. Technol. B* **14**, 1837 (1996).
51. C.-Y. Kim, K.-Y. Kim, J.-S. Roh, I.-C. Ryu, S.-G. Jin, N.-J. Kwak, and S.-K. Lee, VMIC Proceedings, p. 121, 1998.
52. S.-Y. Tai, C.-S. Huang, C. Yi, S. Chang, and R. Tang, VMIC Proceedings, p. 133, 1998.
53. I.-H. Lee, C.-M. Jeong, J.-S. Kim, P.-G. Shon, K.-R. Yoon, K.-K. Son, and S.-K. Lee, VMIC Proceedings, p. 141, 1998.
54. H. Kheyrandish, J. S. Colligon, and J.-K. Kim, *J. Vac. Sci. Technol. A* **12**, 2723 (1994).
55. F. Cerio, J. Drewery, E. Huang, and G. Reynolds, AVS Symposium, San Jose, Paper No. ThM11, 1997.
56. Y. Tanaka, E. Kim, J. Forster, and Z. Xu, *J. Vac. Sci. Technol., B* **17**, 416 (1999).
57. Y.-W. Kim, J. Moser, I. Petrov, J. E. Greene, and S. M. Rossnagel, *J. Vac. Sci. Technol. A* **12**, 3169 (1994).
58. Y.-W. Kim, I. Petrov, J. E. Greene, and S. M. Rossnagel, *J. Vac. Sci. Technol. A* **14**, 346 (1996).
59. J. Su, Z. Wu, G. Yao, C. Cha, M. Abburi, M. Narasimhan, and Z. Xu, VMIC Proceedings of 124, 1998.
60. J. Musil, V. Poulek, V. Valvoda, R. Kuzel, Jr., H. A. Jehn, and M. E. Baumgartner, *Surf. Coating Technol.* **60**, 484 (1993).
61. Y. Tanaka, G. D. Yao, J. VanGogh, B. Herner, J. Y. Zhang, H.-G. Wang, L. Buckley, P. Chakravarthy, A. Mak, and S. Ghanayem, *Proceedings of the 5th ICSITC Conference*, p. 207, IEEE Press, New York, 1998.
62. G. A. Dixit, W. Y. Hsu, A. J. Konecni, S. Krishnan, J. D. Luttmer, R. H. Havemann, J. Forster, G. D. Yao, M. Narasimhan, Z. Xu, S. Ramaswami, F. S. Chen, and J. Nulman, *IEDM Tech. Dig.*, 357 (1996).
63. H. J. Barth, H. Helneder, D. Piscevic, M. Scneegans, G. Birkmaier, G. Crowley, H. Kieu, S. Ramaswami, and U. Richter, VMIC Proceedings, p. 225, 1997.
64. S. Bothra, S. Sengupta, B. Chang, M. Narasimhan, and S. Ramaswami, VMIC Proceedings, p. 240, 1997.
65. Z. Wang, W. Catabay, J. Yuan, J. Ku, N. Krishna, V. Pavate, A. Sundarajan, D. Saigal, B. Chang, M. Narasimhan, J. Egermeier, and S. Ramaswami, VMIC Proceedings, p. 258, 1997.
66. Z. Xu, K. Ngan, J. V. Gogh, R. Moseley, Y. Tanaka, H. Kieu, F. Chen, and I. Raaijmakers, *SPIE Conf.* Microelectronics Technol. Process Integration **2335**, 70 (1994).
67. B. Chang, S. Hui, C. Cha, S. Lee, M. Nam, E. Kim, H. Kieu, K. Ngan, G. Yao, Z. Xu, and F. Chen, VMIC Proceedings, p. 389, 1997.
68. B. Chang, personal communication, 1997.
69. V. M. Dubin, C. H. Ting, and R. Cheung, VMIC Proceedings, p. 69, 1997.
70. P. Ding, T. Chiang, B. Sun, R. Tao, I. Hashim, L. Chen, G. Yao, B. Chin, R. Moseley, Z. Xu, and F. Chen, VMIC Proceedings, p. 87, 1997.
71. B. Zhao, D. Feiler, V. Ramanathan, Q.-Z. Liu, M. Brongo, J. Wu, H. Zhang, J. C. Kuei,

D. Young, J. Brown, C. Vo, W. Xia, C. Y. Chu, J. Zhou, L. Tsau, D. Dornish, L. Camilletti, P. Ding, G. Lai, B. Chin, M. Johnson, J. Turner, T. Ritzdorf, G. Wu, and L. Cook, *Symp. VLSI Technol. Digest*, in press.

72. D. Li, Y. W. Chung, S. Lopez, M. S. Wong, and W. D. Sproul, AVS Symposium, Paper No. TF-WeP15, 1994.

Plasma Physics

JEFFREY A. HOPWOOD

Northeastern University, Boston, Massachusetts

I. Introduction

In this chapter a basic physical model of ionized physical vapor deposition (I-PVD) is developed. The goal is to understand the salient mechanisms connecting the externally controlled system parameters to the internal plasma parameters that determine film quality. Since the purpose is to foster an intuitive understanding of I-PVD, the physical descriptions will be simple analytical models that successfully describe experimentally observed results. A more detailed numerical simulation can be found in Chapter 8.

Many I-PVD reactor configurations have been designed and characterized. In this chapter, generic principles of operation will be discussed that can be applied to many of the various reactors. As shown in Fig. 1, a cylindrical vacuum chamber with a height *L*, defined by the throw distance between the sputter target and the wafer, is common to all I-PVD sources. Electrical power is supplied to the target, the plasma, and the wafer so that the sputtered metal density, electron density, and ion energy at the wafer surface may be externally controlled. The plasma is typically generated by radio frequency inductive coupling (Fig. 2), microwave electron cyclotron resonance, or a DC hollow cathode magnetron. The precise method of plasma generation is not material to the physical processes as long as an

Vol. 27
ISBN 0-12-533027-8
ISSN 1079-4050/00 $30.00

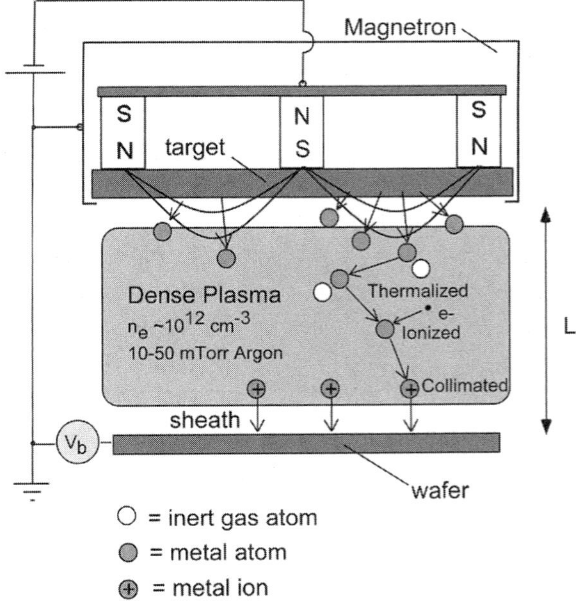

FIG. 1. In I-PVD sputtered metal atoms are collisionally slowed to thermal speeds and ionized by electron impact. The ions that diffuse toward the wafer are collimated and accelerated by the plasma sheath.

electron density of $\sim 10^{12}$ cm^{-3} is created. The pressure within the reactor is in the range of 10–50 mTorr. An inert gas such as argon is used for deposition of pure metals, but nitrogen may be added to the plasma if metal–nitride compounds are needed.

In the following sections, the physics of I-PVD will be described so that the chamber geometry, plasma power, gas pressure, gas type, and target power can be selected to achieve the proper degree of metal ionization, collimation, and uniformity.

II. Ionization Mechanisms

The essence of I-PVD is to physically create a metal vapor flux, ionize the vapor using a high-density plasma, and collimate the ion flux using the plasma's sheath. Initially this vapor flux consists of neutral atoms and is created using conventional methods such as sputtering or evaporation. The high-density plasma is generated using a gas with a significantly

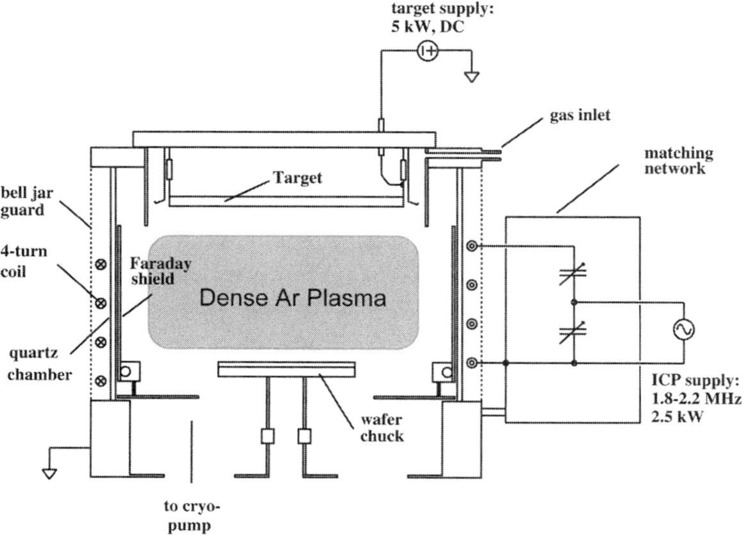

FIG. 2. Cross section of the I-PVD reactor using an inductively coupled plasma (ICP) to ionize sputtered target atoms.

higher ionization potential than that of the metal atoms. For example, copper atoms ($E_{\mathrm{IP}} = 7.73\,\mathrm{eV}$) are readily ionized in a plasma of argon ($E_{\mathrm{IP}} = 15.76\,\mathrm{eV}$). If the argon density is much greater than the copper vapor density, the copper is highly ionized by electron impact since the electron temperature (T_e) of the plasma is determined primarily by the ionization potential of argon.

A. IONIZATION DISTANCE

The average distance that a metal atom travels through a plasma before being ionized is an important quantity since ultimately this determines the geometry of the I-PVD source. If we assume that the metal atom (M) will be ionized by a single electron impact,

$$M + e^- \xrightarrow{K_\mathrm{i}} M^+ + 2e^-, \tag{1}$$

then the mean ionization path length (λ_{iz}) can be determined from

$$\lambda_{\mathrm{iz}} = \frac{v_M}{n_e K_\mathrm{i}(T_e)}, \tag{2}$$

where v_M is the velocity of the metal atom, $K_i(T_e)$ is the ionization rate constant from Eq. (1), and n_e is the electron density within the plasma. The ionization rate constant can be expressed in the Arrhenius form:

$$K_i(T_e) = K_0 \exp(-E_0/T_e), \tag{3}$$

where $K_0 = 5.6 \times 10^{-8}\,\text{cm}^{+3}\,\text{sec}^{-1}$ and $E_0 = 8.8\,\text{eV}$ for copper ionization. From Eq. (2) it is clear that fast metal atoms require longer distances to ionize. Higher electron density and higher electron temperature are desirable, however, since these plasma properties decrease the mean ionization path length.

Sputtered atoms exhibit a Thomson distribution[1] in which the most probable energy is one-half the surface binding energy ($\sim 2\,\text{eV}$ for Cu). Therefore, $\lambda_{iz}(\text{Cu}) \sim 80\,\text{cm}$ in a typical Ar plasma, where $T_e = 3\,\text{eV}$ and $n_e = 10^{12}\,\text{cm}^{-3}$. It is impractical to construct such a large I-PVD tool and to produce an 80-cm-long, high-density plasma. The alternatives to large reactors are (i) use evaporation to create a less energetic metal flux[2] or (ii) use a high background pressure of inert gas so that the sputtered atoms are thermalized.[3,4] Since sputtering is considered to be the more desirable PVD technology, the second option is currently the most commonly practiced in I-PVD. Once the sputtered metal flux is collisionally cooled, the typical ionization path length is reduced to $\sim 10\,\text{cm}$. I-PVD accomplished through thermalization, ionization, and collimation as shown schematically in Fig. 1 will be the focus of the remainder of this chapter.

B. METAL IONIZATION MODEL

Here, a zero-dimensional, spatially averaged model for metal–argon plasmas[5] is presented. This model provides insight into the dominant ionization mechanisms occurring within the I-PVD plasma. The model balances generation rates for argon excited states (Ar*), argon ions (Ar^+), and metal ions (M^+) with the radiative and diffusive loss rates of these plasma species. For a given gas pressure and electron density, the plasma's electron temperature and metal ion fraction will be self-consistently determined.

The model assumes that the sputtered metal is thermalized by collisions with the argon background gas as described previously. The three most important ionizing collisions responsible for the generation of M^+ in an inert gas plasma are electron impact ionization of the metal neutral, electron impact ionization of an excited metal atom, and Penning ionization by collision with an electronically excited argon atom. The specific collisions with the metal atom, M, are

$$M + e^- \xrightarrow{K_i} M^+ + 2e^- \tag{4}$$

$$M^* + e^- \xrightarrow{K_i^*} M^+ + 2e^- \tag{5}$$

$$\text{Ar}^* + M \xrightarrow{K_p} M^+ + \text{Ar} + e^-, \tag{6}$$

where K_i, K_i^*, and K_p are the rate constants for the two electron impact collisions and Penning ionization, respectively.

To determine the diffusive loss rate of M^+, it is observed that I-PVD processes typically occur at gas pressures of 10–50 mTorr. In addition, a cylindrical chamber with a radius of approximately twice the wafer radius, e.g., $R \sim 20$ cm, is used. The ratio of the ion density at the edge of the plasma to the ion density at the center (n_{io})[6,7] is

$$h_R = \frac{n_i(r = R)}{n_{io}} \approx \frac{0.8}{\sqrt{4 + (R/\lambda_i)}} = 0.06, \tag{7}$$

where λ_i is the ion-mean free path. Note that $\lambda_i(\text{cm}) \approx 3/p$, where p is the pressure in mTorr. The ion mean free path is distinct from λ_{iz} since it includes nonionizing collisions, primarily with neutral Ar. Since the ion density near the wall of the chamber is quite small (see Eq. 7), the solution to the source-free ambipolar diffusion equation

$$\frac{\partial n_i}{\partial t} = D_a \nabla^2 n_i \tag{8}$$

with the boundary condition that $n_i(r = R) = n_i(z = 0) = n_i(z = L) = 0$ is used to determine the ion lifetime (τ_i) in the plasma,[8]

$$\frac{1}{\tau_i} = D_a \left[\left(\frac{\pi}{L}\right)^2 + \left(\frac{2.405}{R}\right)^2 \right], \tag{9}$$

where $D_a = k_B T_e \mu_+ / e$ is the ambipolar diffusion coefficient of M^+ in an inert gas and L is the distance between the wafer and the target. The mobility (μ_+) of various metal ions in Ar is estimated using the Langevin formula in the polarization limit.[9] Setting the total generation rate of metal ions (Eqs. 4–6) equal to the diffusive loss rate (Eq. 9), one finds the fundamental particle balance equation for M^+:

$$K_p n_{\text{Ar}^*} n_M + K_i n_e n_M + K_i^* n_e n_{M^*} = \tau_{i(M^+)}^{-1} n_{M^+}, \tag{10}$$

where n_j is the density of the plasma species indicated by the subscript j.

Since the rate constants for ionization of excited metal species (M^*) are not well characterized, K_i^* will be ignored. This approximation will result in an underestimate of the total metal ion production. The Penning ionization rate constant is determined from $K_p = \sigma_p v_{th}$, where σ_p is the Penning cross section and v_{th} is the thermal velocity. In the absence of experimentally measured Penning cross sections for metals of interest, published data[10] for Zn and Cd will be scaled by the square of the atomic radii. This is a crude approximation to the actual cross sections, but, as will be shown, the ionization of metal in a high-density argon discharge is primarily due to electron impact ionization. Relatively large errors in the Penning cross section, therefore, have little effect on the calculated ionization of metals in I-PVD.

The ionization of M by the Penning process depends on the excited argon density n_{Ar^*} which is produced by electron collisions

$$\mathrm{Ar} + e^- \xrightarrow{K_e} \mathrm{Ar}^* + e^-. \tag{11}$$

The loss rate of Ar* depends on the deexcitation by Penning ionization (Eq. 6) and diffusion losses. The diffusion lifetime (τ_M) is determined by free diffusion of long-lived metastable states from the plasma to the chamber walls where deexcitation occurs. The lifetime is similar to Eq. 9:

$$\frac{1}{\tau_m} = D_m\left[\left(\frac{\pi}{L}\right)^2 + \left(\frac{2.405}{R}\right)^2\right], \tag{12}$$

where D_m is the metastable diffusion coefficient.[11] Other loss mechanisms for Ar* are imprisoned resonant radiative decay,[12] which will be denoted by the lifetime τ_r, and collisional ionization of the excited states

$$\mathrm{Ar}^* + e^- \xrightarrow{K_c} \mathrm{Ar}^+ + 2e^-. \tag{13}$$

The particle balance equation for the generation and loss of excited argon is

$$K_e n_e n_{Ar} = n_{Ar^*}\{K_c n_e + K_p n_M + \tau_m^{-1} + \tau_r^{-1}\} \equiv \tau^{-1} n_{Ar^*}. \tag{14}$$

Noting that the total metal density is $n_o = n_M + n_{M^+}$, the metal ion fraction in a plasma with electron density n_e is found from Eqs. (10) and (14) to be

$$\frac{n_{M^+}}{n_0} = \frac{(K_i + K_p K_e \tau n_{Ar})\tau_i n_e}{1 + (K_i + K_p K_e \tau n_{Ar})\tau_i n_e}. \tag{15}$$

To solve Eq. (15) requires a knowledge of the electron temperature since the electron-impact rate constants are determined from

$$K_j(T_e) = \langle \sigma_j v \rangle = \int_0^\infty v\sigma_j(E) f(T_e, E)\, dE, \tag{16}$$

where σ_j are the collision cross sections for the various reactions discussed previously, v is the electron velocity, and $f(T_e, E)$ is the electron energy distribution function. To hasten convergence to the solution, the rate constants are expressed in an Arrhenius form by integrating cross sections from the literature using a Maxwellian distribution in Eq. (16) and numerically fitting the results to

$$K_j(T_e) = K_0 \exp(-E_0/T_e). \tag{17}$$

Table 1 lists the constants K_0 and E_0 used in this model and references to the original cross-section data. Also included is the mobility of the species produced by each collision at standard temperature and pressure (STP).

The electron temperature can be self-consistently determined by adding a quasi-neutrality requirement to the previous set of equations, $n_e = n_{Ar^+} + n_{M^+}$. This requires a calculation of the argon ion density from the particle balance for Ar^+:

$$K_a n_{Ar} n_e + K_c n_{Ar*} n_e = \tau_{i(Ar^+)}^{-1} n_{Ar^+}, \tag{18}$$

where the rate constant K_a represents the single-step ionization process

$$Ar + e^- \xrightarrow{K_a} Ar^+ + 2e^-. \tag{19}$$

The fraction of ionized metal may be iteratively determined from the expressions outlined previously for a given argon gas density and temperature (pressure), metal density (target power), electron density (plasma power), and chamber geometry (R, L). The rate constants are calculated beginning with an initial guess for the electron temperature. Equations (15) and (18) are then used to find the metal ion and argon ion densities. The

TABLE 1
RATE CONSTANT AND MOBILITY FOR SPECIES IN I-PVD PLASMAS

	$K_0 \times 10^{-8}$ (cm^3/sec)	E_0 (eV)	Reference	μ_0 at STP (cm^2 v^{-1} sec^{-1})
K_e: Ar→Ar*	2.2	12.4	de Heer *et al.*[13]	2.7[11]
K_c: Ar*→Ar^+	21	5.3	Vriens[14]	1.6[15,16]
K_a: Ar→Ar^+	12.3	18.68	Rapp and Englander-Golden[17]	1.6
K_i: Al→Al^+	12.3	7.23	Shimon *et al.*[18]	2.7[9]
K_i Ti→Ti^+	23.4	7.25	Lennon *et al.*[19]	2.3[9]
K_i: Cu→Cu^+	5.62	8.77	Lotz[20]	2.2[9]
K_i: C→C^+	4.0	12.6	Lotz[20]	3.6[9]

sum of these two positive ion densities is compared with the chosen electron density, and the electron temperature is increased or decreased until charge neutrality is achieved.

C. IONIZATION MODEL RESULTS

1. *Conversion of Metal Density to Flux*

From the perspective of film deposition, one is interested in the fluxes of depositing species. The model outlined previously, however, provides only the densities of species in the bulk of the plasma. To convert metal ion density to metal ion flux at a surface (Γ_i) the acceleration of ions to the Bohm velocity $(k_B T_e/m_i)^{1/2}$ by the plasma's presheath must be accounted for:

$$\Gamma_i = 0.61\, n_{M^+}(k_B T_e/m_i)^{1/2}, \tag{20}$$

where m_i is the ion mass and k_B is the Boltzmann constant. The flux of thermalized neutral metal, on the other hand, is simply

$$\Gamma_n = \tfrac{1}{4} v_{th} n_M, \tag{21}$$

where $v_{th} = (8\, k_B T_g/\pi m)^{1/2}$ is the mean thermal velocity of the gas. In nonequilbrium discharges T_e is much greater than T_g. This makes the fraction of ionized metal *flux* to a wafer larger than the fraction of ionized metal in the plasma. For example, if the bulk ionization $n_{M^+}/(n_{M^+} + n_M)$ is 0.3, then the fraction of ionized metal flux is $\Gamma_i/(\Gamma_i + \Gamma_n) \sim 0.8$. One concludes that it is not necessary to completely ionize the sputtered metal to create a highly ionized flux at the wafer surface.

2. *Penning Ionization*

One of the advantages of an I-PVD model is that specific physical processes can be turned on or off. The relative importance of electron impact ionization and Penning ionization is investigated by setting K_i or K_p equal to zero. The ionized flux fraction of aluminum due to electron impact ionization only, Penning ionization only, and both collisions together is shown in Fig. 3. If the electron density is low, Penning ionization is responsible for the majority of metal-ionizing collisions. This observation is consistent with the classic paper by Coburn and Kay[21] in which Penning ionization was shown to be responsible for the generation of ions in diode sputtering. I-PVD, however, requires electron densities two or three orders of magnitude higher than diode sputtering. From Fig. 3 it can be seen that

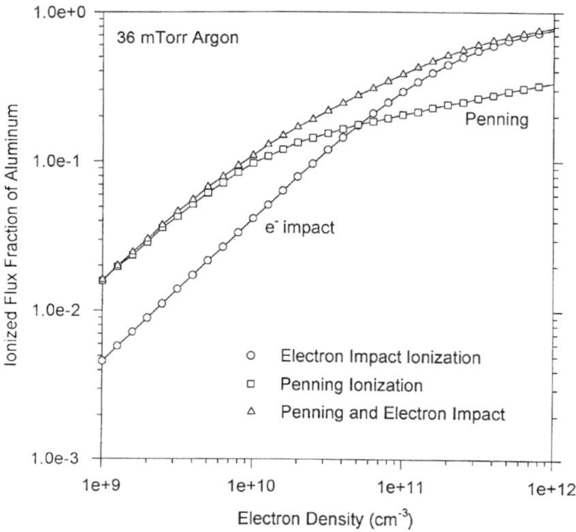

FIG. 3. Electron impact ionization is the primary path for metal ion production in a high electron density plasma. Penning ionization dominates under conditions of low electron density.

when $n_e > 10^{11}\,\text{cm}^{-3}$ most of the metal ions are generated by electron impact. Under I-PVD conditions, the Penning process becomes relatively unimportant. When the electron density is high, metastable Ar is rapidly lost due to electron collisions. Since the density of Ar* increases more slowly than the electron density, Penning ionization becomes marginalized.

3. *Plasma Chamber Size*

Plasma chamber geometry plays an important role in all plasma processes. In addition to controlling plasma uniformity, the chamber size is a prime determiner of electron temperature. As the chamber dimensions decrease, the rate at which ions diffuse to the walls increases. To compensate for an increased loss rate, the electron temperature increases. In the context of I-PVD it is reasonable to expect that the ionization of metal would increase in small chambers since ionizing electron collisions would be more probable. Figure 4 shows that the modeled Al ion fraction actually decreases as the chamber's radius is reduced. The lower degree of ionization is caused by a higher diffusive loss rate of M^+ which overwhelms any benefit derived from higher electron temperature. In general, if the electron density is constant, larger plasma chambers will produce a more highly ionized flux of metal.

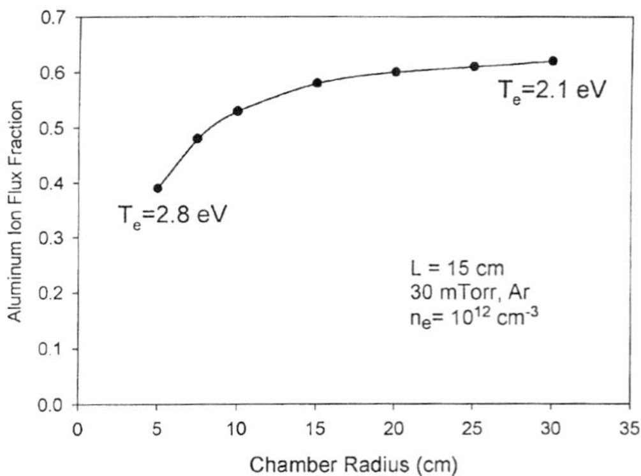

FIG. 4. Although decreasing the I-PVD chamber size increases the plasma's electron temperature, the metal is less ionized due to $M^+ - e^-$ recombination losses at the chamber walls.

4. *Comparison with Measurements*

To verify the validity of the global model, a comparison with experimental measurements is shown in Fig. 5. The elevated gas temperature used in the model ($T_g = 800$ K) is due to energy transferred from the energetic sputtered aluminum atoms to the argon gas during thermalization[22,23] and is discussed later. The error bars for the model show the sensitivity of the calculation to a 20% uncertainty in the measured ionization cross section for aluminum. This comparison shows that the global model predicts the

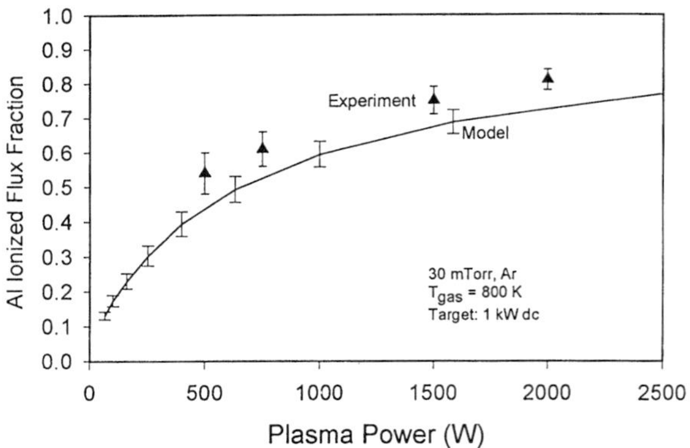

FIG. 5. Comparison of measured and modeled aluminum flux ionization validates the I-PVD model.

ion flux fraction within the inherent inaccuracies of the model and the experiment. One possible reason for the model's consistently underpredicted ionization is that two-step metal ionization

$$M + e \rightarrow M^* + e$$

$$M^* + e \rightarrow M^+ + 2e$$

has been ignored.

5. *Comparison of Various Metals*

Figure 6 shows the modeled ionization of various metals that are either currently used or projected for use in integrated circuit interconnects. The error bars on the Cu data indicate the 30% uncertainty in the ionization cross section. Notice that the ion fraction is a few percent when the electron density is on the order of $10^{10}\ \mathrm{cm}^{-3}$. This is a typical ionization level for conventional sputtering and is primarily due to Penning ionization.[21] At higher electron density, however, the excited-state argon density is inhibited by electron collisions[24] (Eq. 13) such that electron impact ionization dominates the Penning ionization of M. At electron densities greater than $\sim 10^{11}\ \mathrm{cm}^{-3}$, the important parameters that determine how completely a metal flux will be ionized in I-PVD are the electron impact ionization cross section, ionization potential, and the mobility of M^+ in argon. The first two factors relate to the generation of metal ions, whereas the ion mobility sets the loss rate. For example, although Ti has a higher ionization threshold

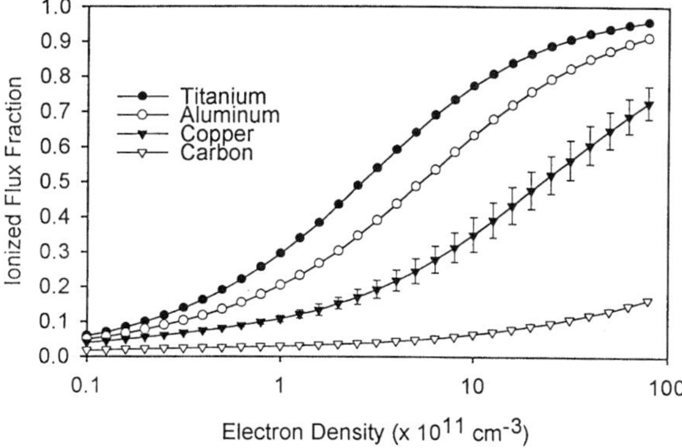

FIG. 6. A comparison of technologically important metals shows that low ion mobility and a large ionization cross section result in the highest degree of ionization. Carbon is difficult to ionize since its cross section is small and the C^+ is highly mobile in argon (see Table 1).

than Al, the relative ionization of Ti atoms is higher since the cross section is twice that of Al while the mobility of Ti^+ is slightly lower (see Table 1). Copper, on the other hand, has a much lower cross section and a higher ionization potential than those of Ti or Al. These factors make Cu more difficult to ionize. Finally, although not related to interconnects, carbon ionization is included in the plot to demonstrate that atoms with high ionization potential (11.26 eV for C) and high ion mobility are not significantly ionized in an argon plasma even at densities approaching $n_e = 10^{13}\,\mathrm{cm}^{-3}$.

6. *Background Gas Type*

Although argon is the most common background gas used in I-PVD, the other noble gases are worth discussing. Since metal ionization occurs primarily through electron impact ionization, it is advantageous to choose a background gas that supports a high electron temperature. Electron temperature in the plasma increases with the ionization potential of the gas. The more massive inert gases such as Kr and Xe have low ionization potentials and are inefficient ionizers of most metal atoms. It is therefore reasonable to investigate neon ($E_{IP} = 21.56$ eV) and helium ($E_{IP} = 24.59$ eV) as possible substitutes for argon ($E_{IP} = 15.76$ eV).

Figure 7 shows a comparison of ionization fraction between argon and neon plasmas. At low electron density ($10^{10}\,\mathrm{cm}^{-3}$), the ionization occurs by

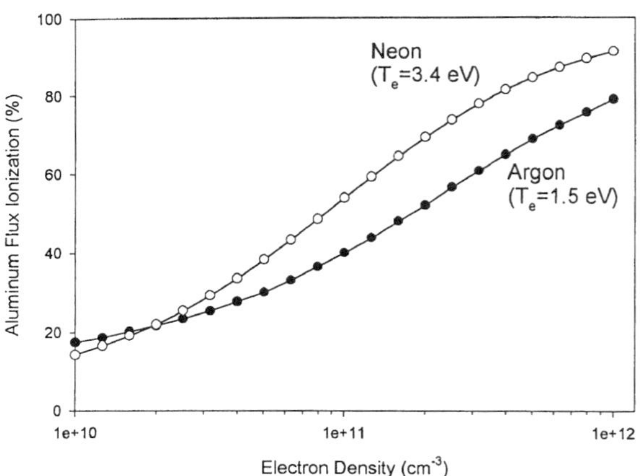

FIG. 7. A comparison of Ar and Ne plasmas shows that neon produces both a higher electron temperature and a more highly ionized flux of aluminum.

the Penning process and argon provides more metal ionization since its Penning cross section is larger than that of neon. At higher electron density the electron impact ionization collisions dominate and one finds that Ne produces a hotter electron gas and a larger fraction of Al^+. This comparison is made for Ne and Ar plasmas of equal electron density. In practice, more plasma power will be needed if Ne is to produce the same electron density as Ar.

Although helium plasmas will produce an even more energetic electron gas than neon, the ion fraction of metal is generally lower. The difficulty encountered with He is very rapid diffusion of metal ions and electrons through the helium background to the chamber walls. Despite the high ionization frequency, the increased loss rate of M^+ results in a lower ion fraction.

Finally, it must be remembered that the inert background gas plays roles other than the generation of a hot, dense electron gas. The ions from the inert gas bombard the metal target and sputter atoms into the plasma region. Very light ions, such as Ne and He, exhibit low sputter yields and therefore do not produce many sputtered metal atoms per incident ion. The other important function of the background gas is to thermalize the energetic sputtered atoms. Light noble gas atoms are poor absorbers of this energy due to the mismatch of mass with heavier metal atoms. This generally means that the lighter noble gases allow a larger fraction of the sputtered flux to traverse the plasma region at high velocity. These fast metal atoms are unlikely to be ionized and are deposited at the wafer as uncollimated neutrals. For these reasons argon is the most commonly used gas in I-PVD.

7. *High Metal Vapor Density*

As metal vapor is added to an argon plasma, the electron temperature cools due to the low ionization and excitation potentials of the metal atoms. Figure 8 shows this cooling phenomena as calculated from the global model. For this calculation the electron energy distribution function was assumed to be Maxwellian. In reality the high-energy tail of the distribution function will probably be depleted by inelastic collisions with metal species. In either case, the loss of high-energy electrons ($E > E_{IP}$) results in a decreased average electron energy. Lower energy electrons are less likely to ionize metal atoms. The end result is a decrease in the ionization probability of metal atoms as the flux of metal vapor is increased. Ultimately, this will limit the ion fraction attainable under high-rate deposition conditions, i.e., high target power.

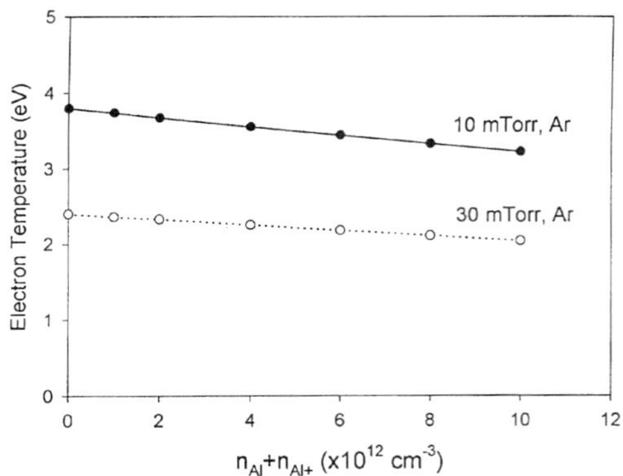

FIG. 8. A high density of metal atoms, created by a high target power, cools the electron temperature in the I-PVD plasma if n_{Ar} is constant.

8. *Gas Rarefaction*

The argument for electron temperature quenching given previously is slightly oversimplified. Although high fluxes of sputtered metal atoms will decrease the average electron energy if the Ar density is constant, the sputtered metal flux will also simultaneously decrease the argon density. The decrease in argon density will have the opposite effect of metal-induced quenching. In other words, decreased gas density will allow electrons and ions to diffuse to the chamber walls more quickly and the electron temperature will increase to offset the enhanced loss rate.

Gas rarefaction is a well-known phenomenon in diode sputtering. The average energy of a sputtered atom is $E_S \sim 10\,\text{eV}$. As the sputtered atom is thermalized by collisions with the background gas, the temperature of the gas increases. Since most vacuum systems maintain a constant pressure and volume, the transfer of energy from sputtered particles to the Ar gas decreases n_{Ar}. In conventional magnetron sputtering the rarefaction is minimal since the throw distance is short ($\sim 5\,\text{cm}$) and the pressure is low ($\sim 2\,\text{mTorr}$) so that the sputtered neutral's mean free path is greater than the throw distance. Unfortunately, in I-PVD it is necessary to thermalize the sputtered species in order to increase the probability of ionization. Thermal considerations, therefore, cannot be ignored in I-PVD.

A simple model for gas heating due to sputtering[23] assumes that the chamber wall temperature is fixed and that the heat flux from the region below the target is spherically symmetric. The power deposited by thermalized sputtered atoms (P_S) diffuses to the chamber walls such that the gas temperature (T_g) at the center of the plasma is given by

$$T_g - T_{\text{chamber wall}} \approx \frac{P_S}{4\pi K}\left(\frac{1}{r_{\text{th}}} - \frac{1}{R_w}\right), \tag{22}$$

where R_w is the radius of the chamber wall, r_{th} is the distance over which the sputtered atom thermalizes,[25] and K is the thermal conductivity of the gas. The power is $P_S = I_T Y_V E_S/q$, where I_T is the target current and Y_V is the sputter yield. A crude estimate of the thermalization distance is $r_{\text{th}}(\text{cm}) \sim 0.024/p$, where p is the pressure in Torr. Typically, the average energy per sputtered Cu atom[26] is 8.8 eV/atom and the thermal conductivity of argon is 0.019 W m^{-1}K^{-1} at room temperature. From Eq. (22) it can be seen that the argon gas temperature may be readily heated to $T_g > 1000$ K.

There are three main physical changes caused by argon rarefaction. First, the reduction of gas density by energetic sputtered atoms will increase the electron temperature since the loss rate of charged particles through the rarefied background will increase. Depending on the pressure and source geometry (r_{th} and R_w), this effect may partially or completely negate the cooling of the average electron energy due to the metal vapor. Second, for a fixed plasma power the electron density will decrease when more metal is sputtered into the plasma[22] as shown in Fig. 9. The decrease in n_e can be attributed to the increased loss rate of electrons and ions from the hot center of the plasma. Finally, the decrease in argon gas density will allow a greater fraction of the sputtered species to semiballistically traverse the high-density plasma region. The thermalization of the sputtered species becomes less efficient. The latter two consequences of gas heating cause the ionization fraction of the sputtered metal to decrease as the amount of sputtered metal increases. Experimentally,[3,4] the quenching of ion fraction is shown in Fig. 10, in which 80% of the sputtered aluminum flux is ionized when the target is sputtered using 1-kW DC. The ionization drops to only 40%, however, when the amount of sputtered metal is tripled by increasing the target power to 3 kW. This quenching can also be observed in the modeled results when heating of the argon gas is included in the calculation as shown in Fig. 11.

The following are possible solutions to the negative effects of gas heating: (i) Increase the pumping speed and gas flow to shorten the residence time of gas in the chamber, (ii) alter the chamber geometry so that $R_w \sim r_{\text{th}}$ and heat is removed through the chamber walls, or (iii) follow the suggestions of Rossnagel and Joo to pulse the target power, leaving sufficient time for the gas to cool between pulses.

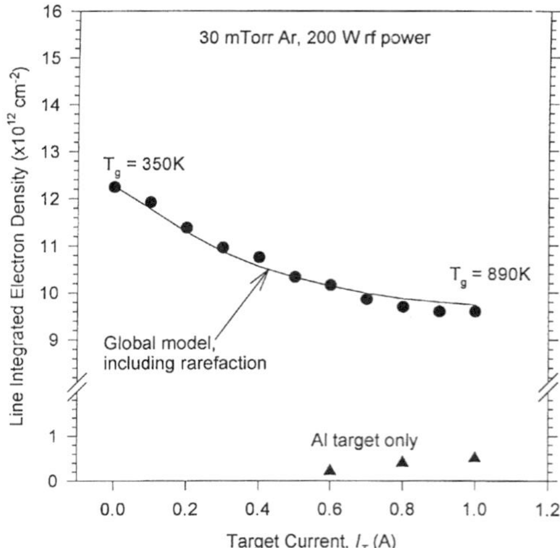

FIG. 9. Experimentally, the electron density generated by an ICP is found to decrease as more metal atoms are sputtered into an argon plasma (•). The solid line is calculated using a global model that includes rarefaction of the argon gas as T_g increases from 350 to 890 K. The electron density exclusively due to the sputtering target (magnetron) is negligible (Δ).

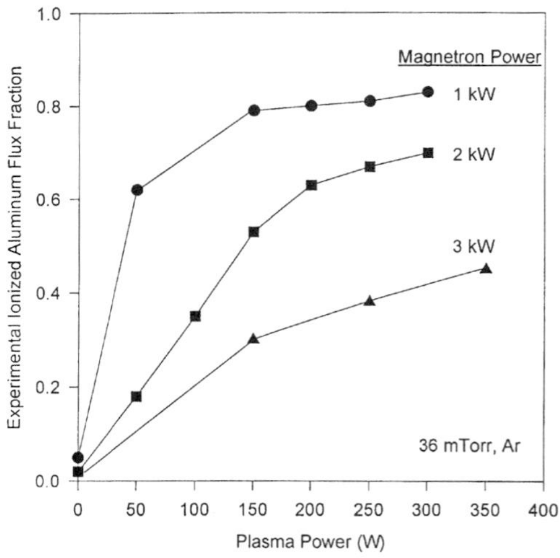

FIG. 10. Experimental data show that increasing the metal density (target power) causes the ion fraction to decrease. This effect is caused by argon rarefaction and electron energy quenching.

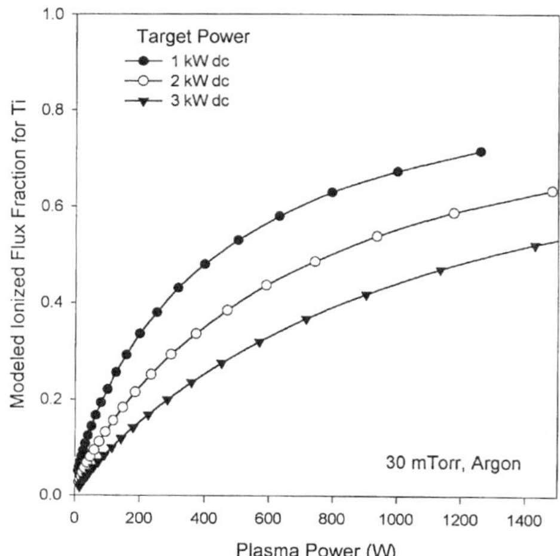

FIG. 11. The global model[22] predicts the observed reduction in ionization due to higher target power.

III. Angular Distribution of Ions

So far, the thermalization and ionization of metal atoms has been discussed. The final step in the I-PVD process is the collimation of metal ions by the plasma sheath prior to deposition on the wafer. This is an important process to understand and control since the ability of I-PVD to deposit materials into high aspect ratio trenches and vias depends on the degree of collimation of the metal ions.

The plasma sheath is actually a two-layer structure consisting of the presheath region and the sheath proper as seen in Fig. 12. The sheath is a boundary layer that supports a strong, perpendicular electric field adjacent to all solid surfaces in the plasma. The sheath above the wafer is important to this discussion since it is responsible for collimating ions from the plasma. The thickness of the sheath (s) can be determined by the Child law[27] if the plasma Debye length and electron temperature (in volts) are known:

$$s = \frac{\sqrt{2}}{3}\lambda_{\mathrm{De}}\left(\frac{2V_0}{T_e}\right)^{3/4} \tag{23}$$

and

$$\lambda_{\mathrm{De}} = 740\sqrt{\frac{T_e}{n_e}}\,\mathrm{cm}, \tag{24}$$

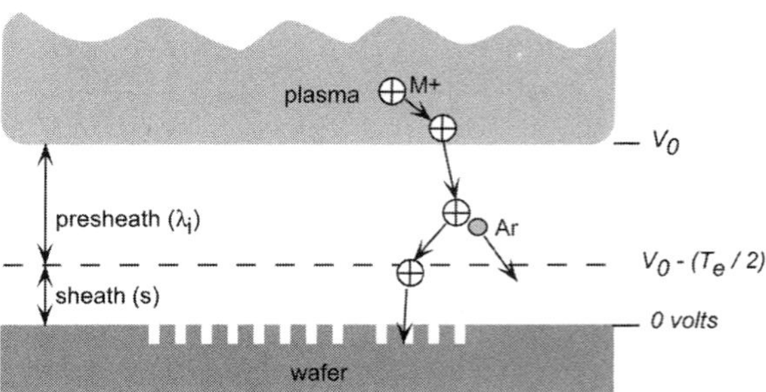

FIG. 12. A detailed view of the plasma–wafer interface reveals a slightly collisional presheath and a collisionless sheath. Collimation of the ions is limited by the initial velocity of the ions in the plasma and collisions in the presheath.

where V_0 is the wafer voltage and n_e is the electron density in cm^{-3}. For low wafer voltages the sheath thickness is several Debye lengths or $\sim 100\,\mu\text{m}$ in I-PVD. Since the ion mean free path is several millimeters, the ions transit the sheath without collisions and may become well collimated.

The presheath region is the layer in which the plasma transitions from a charge-neutral bulk plasma to an ion-dominated sheath. The presheath thickness is approximately one ion mean free path (λ_i).[28] This means that the presheath region will be slightly collisional. The voltage drop across the presheath is $T_e/2$. Ions that are accelerated in the presheath and then scattered by collisions will attain significant velocity *transverse* to the wafer surface and will not be completely collimated by the sheath.

Finally, the collimation of metal ions depends on the initial ion velocity distribution characteristic of the bulk plasma. As the metal atoms are sputtered from the target, the angular distribution is approximately $\cos(\theta)$. Since the average energy of a sputtered atom is $\sim 10\,\text{eV}$, the transverse velocity of an unthermalized metal atom is comparable to the velocity attained in the sheath ($V_0 \sim 15\,\text{eV}$). The metal atoms must be thermalized to reduce this velocity component prior to collimation by the sheath.

The angular distribution of thermal metal ions at the wafer has been determined experimentally[29] by deposition of ionized titanium into high aspect ratio vias. By measuring the film thickness at the bottom of the via relative to the top of the via both the degree of ionization and the effective transverse ion temperature (T_t) are found as shown in Table 2. The effective ion temperature is between 0.13 and 0.18 eV and increases with both pressure and plasma power. The elevated ion temperature is due to two factors. First, the metal atoms are only thermalized to the rarefied argon gas temperature prior to ionization. In addition, the metal ions have a signifi-

TABLE 2
EFFECTIVE TRANSVERSE ION TEMPERATURE T_t(eV) AND ION FRACTION FOR TITANIUM IMPINGING ON A WAFER SURFACE AS A FUNCTION OF ARGON PRESSURE AND PLASMA POWER[a]

	1 kW	2 kW
10 mTorr	0.13 eV, 50%	0.17 eV, 75%
30 mTorr	0.15 eV, 70%	0.18 eV, 85%

[a]The target power was 1 kW for all measurements.

cant probability for an elastic collision in the presheath. This T_t represents a significant reduction from the initial average energy of the sputtered Ti atom (~ 10 eV), however.

Transverse ion temperature results in a divergence angle of the depositing species that limits the bottom coverage of high aspect ratio features. The previous data correspond to a divergence angle of 3–5° from perfect collimation. A simple method of narrowing the angular distribution of the metal ion flux is to apply a negative bias to the wafer such that V_0/T_t is greater. Although the addition of bias narrows the angular distribution of ions, it will also change the properties of the thin film (e.g., film stress). It is more difficult to reduce T_t by additional thermalization of the sputtered species since the argon gas is already quite hot. The gas is typically heated to ~ 0.1 eV from the thermalization of sputtered metal atoms, and any further reduction of T_t will depend on decreasing the argon gas temperature as discussed in the previous section.

IV. Metal Density Distributions

To this point, the I-PVD plasma has been treated as a spatially uniform or volume-averaged discharge. At low gas pressure when diffusion processes are rapid, this is a reasonable approximation. A more detailed physical picture of the I-PVD system, however, must include the axial and radial variation of the metal ion and neutral densities. In the axial direction, an understanding of the evolution of metal ionization from fast neutrals near the face of the target to thermalized ions at the wafer is necessary for determination of an optimized reactor throw distance. Radial uniformity of both the deposition rate and the ionization fraction must also be maintained over 200-mm (and eventually 300-mm) wafers. Design of the reactor for radial uniformity relies on a good physical understanding of the I-PVD diffusion processes. In this section the experimental axial and radial distribution of metal ions and neutrals will be described. Based on these measurements, simple models that are useful for reactor design will be presented.

A. AXIAL DENSITY DISTRIBUTION

The density of metal atoms decreases with an exponential axial dependence as shown in Fig. 13 when the argon background pressure is 30 mTorr.[30] This observation is correct when the sputtered metal atoms are thermalized within a short distance of the target such that the transport of metal atoms becomes diffusive rather than ballistic. In conventional magnetron sputtering the density is more uniform since the pressure is an order of magnitude lower and the energetic metal atoms move with very few collisions to the wafer. Once diffusion dominates at higher pressure (>10 mTorr), many of the metal atoms are lost radially to the chamber sidewalls. Unfortunately, this results in a low efficiency of target usage since much of the sputtered metal is deposited at the reactor wall rather than on the wafer. It is advantageous to minimize the thrown distance such that the metal atoms have sufficient distance to be thermalized and ionized without excessive radial diffusion loss.

Metal ions are distributed much more uniformly than the metal neutrals as shown in Fig. 13. The metal ions are generated by electron impact and Penning collisions throughout the plasma volume, but the metal neutrals only originate from the target. The high degree of metal ionization generated

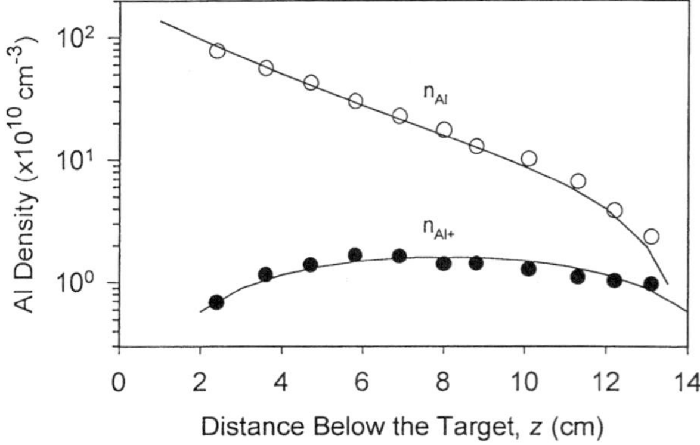

FIG. 13. The density of thermalized aluminum atoms decreases exponentially away from the sputter target. Aluminum ions, on the other hand, are generated throughout the I-PVD reactor. The highest ion fraction occurs at the position downstream where most of the neutral atoms have been depleted by radial diffusion.

by I-PVD is due to the rapid decay of metal neutral density near the wafer and the more or less constant supply of metal ions.

It is apparent that the argon pressure and throw distance of an I-PVD reactor should be chosen such that the neutral density decays from $\sim 10^{12}$ cm^{-3} near the target to the approximate metal ion density ($\sim 10^{10}\ \mathrm{cm}^{-3}$) near the wafer. This will make the ionization $n_{M^+}/(n_{M^+} + n_M) \sim 0.5$ and the fraction of ionized metal flux approximately 0.9.

Simple analytical design equations that predict the metal density can be derived by solving the diffusion equation for metal neutrals,

$$\nabla^2 n_M = 0, \tag{25}$$

with the assumption that the metal density (n_M) is zero on the chamber walls and $n_M = N_0$ in the space just below the target where the metal is thermalized. The neutral metal distribution in a cylindrical chamber of radius R and length L is

$$n_{\mathrm{Al}}(r, z) = 2N_0 \frac{b}{R} \sum_{j=0}^{\infty} \frac{J_1(x_{0j} b/R)}{x_{0j} J_1^2(x_{0j})} J_0\left(\frac{x_{0j} r}{R}\right)\left(\frac{\exp(-k_z z) - \exp(k_z(z - 2L))}{1 - \exp(-2k_z L)}\right),$$

$$k_z = \frac{x_{0j}}{R} \qquad \text{and} \qquad J_0(x_{0j}) = 0, \tag{26}$$

where b is the target radius and J_n is the Bessel function of the first kind. Equation (26) is plotted on Fig. 13 to show the close agreement with the measured aluminum density. Along the center axis ($r = 0$), the density decays approximately as $\exp(-2.405\, z/R)$.

A convenient analytical expression for the metal ion density (n_{M^+}) is given by the Klyarfeld approximation[6,7] for diffusion-dominated discharges between planar boundaries:

$$n_{M^+}(z) \approx n_o\left(1 - (1 - h_l)\left(\frac{z - L/2}{L/2}\right)^2\right), \tag{27}$$

where the ratio of central ion density to sheath-edge density along the chamber axis is determined from the ion mean free path (λ_i) by

$$h_l = \frac{n_{M^+}(z = L)}{n_{M^+}(z = L/2)} \approx \frac{0.86}{\sqrt{3 + (L/2\lambda_i)}}. \tag{28}$$

This model for ion density is plotted along with the measurement of $n_{M^+}(z)$ in Fig. 13. Compared to the model, the actual metal ion density is slightly elevated near the target and depressed near the wafer since the neutral metal density is larger near the target. This increases the ionization frequency at smaller z. The diffusion of metal ions is quite rapid, however, and much of the asymmetry in ionization rate does not contribute to the nonuniformity of n_{M^+}.

B. Radial Density Distribution

The measured density distributions[31] for Ti and Ti^+ are shown in Fig. 14. These densities were measured just above the surface of a 200-mm wafer placed 150 mm below a 300-mm target. The plasma chamber diameter ($2R$) was 450 mm. Both the Ti neutral and ion density are centrally peaked as one would expect in a diffusion-dominated discharge. Note that although the ratio of ions to neutrals in the plasma is only 0.3, the metal flux is dominated by ions since the ions are extracted from the plasma by the presheath (see Eqs. 20 and 21).

The solid curves superimposed on the experimental data in Fig. 14 are simple analytical diffusion models for the neutral and ion density above the wafer. Equation (26) was used to determine n_{Ti} and this expression accurately predicts the neutral density distribution. The ion density above the wafer is modeled by the cylindrical form of the Klyarfeld approximation:

$$n_{M^+}(r) \approx n_o\left(1 - (1 - h_R)\left(\frac{r}{R}\right)^2\right) \tag{29}$$

$$h_R = \frac{n_{M^+}(r = R)}{n_{M^+}(r = 0)} \approx \frac{0.8}{\sqrt{4 + (R/\lambda_i)}}. \tag{30}$$

This expression for the metal ion density works best when the target diameter is approximately equal to the chamber diameter. This is usually the way in which production-class tools are constructed. If the target size is much smaller than the chamber, the Ti neutral density is quite low near the chamber walls. The low density of Ti means that the ionization frequency of metal atoms is very low near the periphery of the chamber and the Ti^+ density will be less than that predicted by the Klyarfeld model.

C. Two-Dimensional Density Distribution

The density of metal ions and atoms throughout the cylindrical I-PVD chamber[32] is shown in the contour plots of Fig. 15. The ion density is maximum near the center of the chamber, whereas the atom density is peaked near the target. More detailed models usually show that the peak

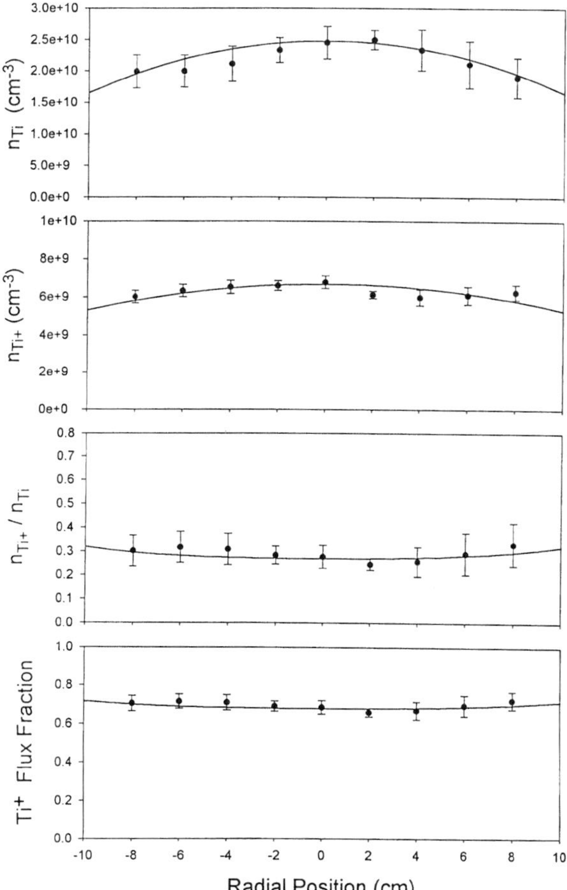

FIG. 14. The radial density distribution of Ti and Ti^+ just above the surface of a 200-mm wafer shows centrally peaked profiles. The diffusion models (1-kW target, 2-kW ICP, 30 mTorr Ar) from the text predict these distributions.

ion density occurs slightly closer to the target than it does in this simple model. The reason for the asymmetry is a higher ionization frequency in the dense metal vapor near the target. In reality, many of the metal atoms near the target are not thermalized and are unlikely to be ionized. Therefore, the assumption of uniform ionization frequency implicit in the Klyarfeld model is more accurate than expected.

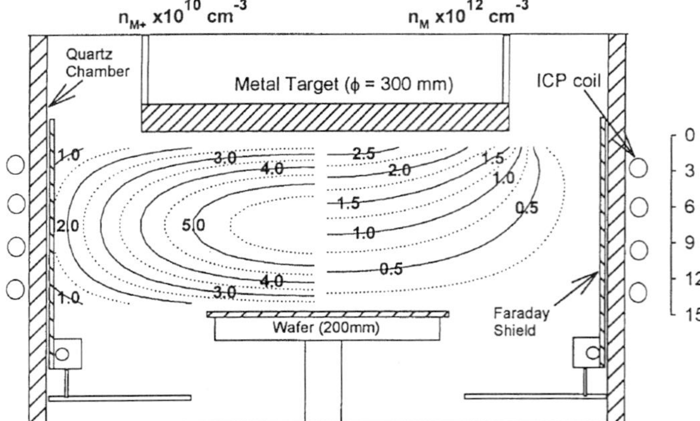

FIG. 15. Contour plots show the metal density distributions within a cylindrical I-PVD reactor calculated from Eqs. (26), (27), and (29).

The contours in Fig. 15 are used to determine the degree of metal ionization throughout the plasma chamber as shown in Fig. 16. Near the target, the plasma is rich in fast metal neutrals and the ion fraction is very low. Downstream, as the metal atoms cool and diffuse to the walls, the metal ion density increases. This produces the peak ion fraction just above the wafer surface. Although both the neutral and ion density are centrally-

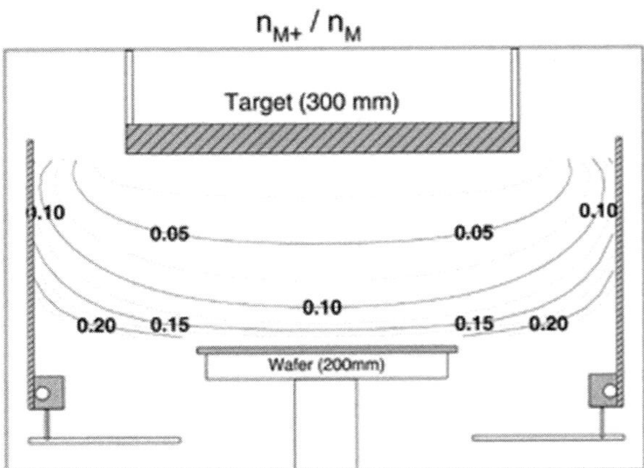

FIG. 16. The metal ionization ratio (n_{M+}/n_M) throughout the bulk of the plasma shows a rapid increase in ionization near the wafer.

peaked and radially nonuniform, the ion fraction at the plane of the wafer is quite uniform since both ion and neutral density distributions are the result of diffusion. This prediction is verified by the experimental data of Fig. 14.

V. Summary

The collimated deposition of sputtered metal typically occurs by a three-step process in I-PVD. First, the fast sputtered atoms are slowed to the thermal velocity by collisions with the background gas. Once the metal is cooled, the probability of ionization by a hot, dense electron gas increases to usable levels. The metal species diffuse to the wafer surface where the ions are accelerated and collimated by the plasma sheath. The directional flux of energetic ions may then be used to deposit thin layers of barrier materials into high aspect ratio features.

The throw distance in I-PVD reactors (L) is somewhat larger than the target-to-wafer spacing in conventional sputtering systems. To ensure thermalization of the sputtered flux, the throw distance should be larger than the thermalization length, which is approximately given by

$$r_{\text{th}}(\text{cm}) \sim 0.024/p, \tag{31}$$

where p is the background pressure in Torr. Thermalization is useful since it greatly increases the probability of ionization, improves collimation by decreasing the transverse velocity of both ions and hot neutrals, and causes unionized metal atoms to diffuse to the chamber walls before depositing on the wafer.

The radial loss of metal atoms to the reactor walls plays an important role in achieving a highly ionized flux of metal to the wafer. Diffusive transport of slowed metal from the thermalization zone near the target toward the wafer results in an exponential decrease in the neutral metal density. Using only the first term in Eq. (26), a quick calculation of the metal neutral decay can be made from

$$n_M(z) \approx N_0[e^{-2.405\, z/R} - e^{2.405(z-2L)/R}]. \tag{32}$$

The metal ion density is $n_{M+} \sim 10^{10}\,\text{cm}^{-3}$ along the central axis when the ionizing plasma provides $n_e \sim 10^{12}\,\text{cm}^{-3}$. To ensure a high ionization fraction of metal flux, the wafer should be positioned at $z = L \gg r_{\text{th}}$ such that the neutral density decays to the metal ion density approximately 1 cm above the wafer, $n_M(L - 1\,\text{cm}) \sim 10^{10}\,\text{cm}^{-3}$. Typically, I-PVD reactors operate in the pressure range of 20–40 mTorr and use a throw distance of $L \sim 10$–15 cm. This thermalizes the sputtered flux within about 1 cm of the target and causes the metal neutral density to decay to $10^{10}\,\mathrm{cm}^{-3}$ in the region just above the wafer.

Since a significant fraction of the sputtered neutrals are lost to the chamber walls in I-PVD, the deposition rate is often quite low. It is important to optimize the throw distance once the operating pressure is known as outlined previously. A crude approximation for currently achieved deposition rates is 10–100 nm/kW of target power. Increased target power will improve the deposition rate, but the ionization of the metal flux will simultaneously decrease. The loss of ion fraction at high target power is due, in part, to dilution by excess sputtered neutrals. An increase in plasma power (i.e., n_e) cannot completely recover the ion fraction, however, because the high flux of energetic sputtered atoms heats and rarefies the argon background gas. The rarefied gas results in a decreased electron density, an increased loss rate of M^+, and incomplete thermalization of the sputtered metal atoms. Care must be taken in the design of the reactor geometry and gas flow to reduce the gas temperature.

Chapter 8 presents a more detailed numerical model of I-PVD operation in which the physical mechanisms discussed here are verified. The numerical approach is also a powerful tool for the examination and optimization of deposition uniformity.

Acknowledgment

This material is based upon work supported by the National Science Foundation under Grant No. DMR-9712988.

References

1. M. W. Thompson, *Philos. Mag.* **18**, 377 (1968).
2. W. M. Holber, J. S. Logan, J. J. Grabarz, J. T. C. Yeh, J. B. O. Caughman, A. Sugerman, and F. E. Turene, *J. Vac. Sci. Technol. A* **11**, 2903 (1993).
3. S. M. Rossnagel and J. Hopwood, *J. Vac. Sci. Technol. B* **12**, 449 (1994).
4. S. M. Rossnagel and J. Hopwood, *Appl. Phys. Lett.* **63**, 3285 (1993).
5. J. Hopwood and F. Qian, *J. Appl. Phys.* **78**(2), 758 (1995).
6. M. A. Lieberman and A. J. Lichtenberg, *Principles of Plasma Discharges and Materials Processing*, p. 140, Wiley, New York, 1994.
7. V. A. Godyak, *Soviet Radio Frequency Discharge Research*, p. 79ff, Delphic, Fall Church, VA, 1986.
8. B. E. Cherrington, *Gaseous Electronics and Gas Lasers*, p. 119, Pergamon, Oxford, UK, 1979.
9. L. M. Chanin and M. A. Biondi, *Phys. Rev.* **107**, 1219 (1957).
10. L. A. Riseberg, W. F. Parks, and L. D. Schearer, *Phys. Rev. A* **8**, 1962 (1973).

11. A. H. Futch and F. A. Grant, *Phys. Rev.* **104**, 356 (1956).
12. T. Holstein, *Phys. Rev.* **83**, 1159 (1951).
13. F. J. de Heer, R. H. Jansen, and W. van der Kaay, *J. Phys. B Atom. Mol. Phys.* **12**, 979 (1979).
14. L. Vriens, *Phys. Lett.* **8**, 260 (1964).
15. M. A. Biondi and L. M. Chanin, *Phys. Rev.* **94**, 910 (1954).
16. K. B. McAfee, D. Siple, and D. Edelson, *Phys. Rev.* **160**, 130 (1967).
17. D. Rapp and P. Englander-Golden, *J. Chem. Phys.* **43**, 1464 (1965).
18. L. L. Shimon, E. I. Nepiipov, and I. P. Zapesochnyi, *Sov. Phys. Tech. Phys.* **20**, 434 (1975).
19. M. A. Lennon, K. L. Bell, H. B. Gilbody, J. G. Hughes, A. E. Kingston, M. J. Murray, and F. J. Smith, *J. Phys. Chem. Ref. Data* **17**, 1285 (1988).
20. W. Lotz, *Z. Physik* **232**, 101 (1970) (and references therein).
21. J. W. Coburn and E. Kay, *Appl. Phys. Lett.* **18**, 435 (1971).
22. M. Dickson, F. Qian, and J. Hopwood, *J. Vac. Sci. Technol. A* **15**(2), 340 (1997).
23. S. M. Rossnagel, *J. Vac. Sci. Technol. A* **6**, 19 (1988).
24. B. E. Cherrington, *Gaseous Electronics and Gas Lasers*, p. 169, Pergamon, Oxford, UK, 1979.
25. A. Gras-Marti and J. A. Valles-Abarca, *J. Appl. Phys.* **52**, 1071 (1983).
26. J. Dembrowski, H. Oechsner, Y. Yamamura, and M. Urbasssek, *Nucl. Instr. Methods B* **18** 464 (1987).
27. M. A. Liebermann and A. J. Lichtenberg, *Principles of Plasma Discharges and Materials Processing*, p. 165, Wiley, New York, 1994.
28. J. A. Meyer, G.-H. Kim, M. J. Goeckner, and N. Hershkowitz, *Plasma Sources Sci. Technol.* **1**, 147 (1992).
29. G. Zhong and J. Hopwood, *J. Vac. Sci. Technol. B* **17**, 405 (1999).
30. M. Dickson and J. Hopwood, *J. Vac. Sci. Technol. A* **15**, 2307 (1997).
31. M. Dickson, G. Zhong, and J. Hopwood, *J. Vac. Sci. Technol. A* **16**, 523 (1998).
32. J. Hopwood, *Phys. Plasma* **5**, 1624 (1998).

Numerical Modeling

MING LI, MICHAEL A. VYVODA, AND DAVID B. GRAVES

Department of Chemical Engineering, University of California at Berkeley, Berkeley, California

I. Introduction

One of the important questions that will determine the effectiveness of ionized physical vapor deposition (I-PVD) tools in practice is how to achieve processing uniformity across the wafer. Perhaps the biggest single motivation behind the development and use of multidimensional simulation of I-PVD is to understand and control uniformity in deposition rate and in other processing characteristics. These include the fraction of the depositing species that are ionic or neutral and, in the case of compound film deposition, the film composition and microstructure.

Simulations are conducted based on a set of model assumptions. We must decide in advance what physics and chemistry to include in the model, then decide how to represent these processes mathematically, and, finally, find some way to numerically solve the equations. Comparisons of simulation predictions to experimental measurements determine the adequacy of the combination of physical model/mathematical representation/numerical sol-

Vol. 27
ISBN 0-12-533027-8

ISSN 1079-4050/00 $30.00

ution. If the model reliability is inadequate, then adjustments must be made in one or more parts of the modeling sequence. On the other hand, if the model is judged to be satisfactory, then it can be used to explore how design or operating variables affect the solution. Model adequacy can only be properly assessed in the context of the engineering (or scientific) problem to be addressed. Usually, there is an iterative interaction between diagnostics and measurements and model/simulation predictions, leading to a deeper understanding of the process. This leads to modifications of the tool design and/or operation to achieve an optimal solution to design and processing objectives.

If our primary objective in simulating I-PVD tools is to better control uniformity, then what physics and chemistry need to be included? First, we assume azimuthal symmetry and a cylindrical geometry. Any three-dimensional (3-D) (θ direction) nonuniformities are ignored. This means that 3-D magnetic fields, often used in magnetron target configurations, must be either ignored or treated in a quasi-2-D way. We can approximately treat the effects of the target radial nonuniformity by simply assuming a target sputtering rate profile that follows the experimentally observed target erosion profile. We further assume that the plasma density is sustained only through the RF inductive power deposition. Target power in this model serves only to accelerate ions to the target for sputtering. Given these assumptions, the I-PVD plasma model closely resembles a conventional ICP model (Bukowski *et al.*[1]). For the model presented here, we have chosen the simplest case of argon working gas and an aluminum target.

The basic physical picture of the I-PVD process that underlies our model can be described as follows. Ar is introduced into the chamber and is pumped in the exhaust. Metal atoms are sputtered from the target with some prescribed radial profile. A high-density plasma containing neutral and ionic metal species and working gas (Ar) species is sustained with immersed metal-sheathed radio frequency (RF) coils. Power is coupled inductively from these coils into electrons, maintaining a plasma density that can be in the $10^{12}\ \mathrm{cm}^{-3}$ range. Capacitive coupling from the coils may be important in causing ions to sputter the metal on the coils. In general, metal deposition from neutral and ionic metal species occurs on all surfaces in contact with the plasma: target, coils, walls, and substrate (wafer). The substrate may be RF biased and capacitively coupled to the plasma.

The study of processing uniformity discussed previously is approximately based on the following phenomena. RF power is inductively deposited locally into electrons. Electron temperature gradients from the point of heating cause gradients in rate coefficients for electron-impact ionization. Positive ions move to walls via ambipolar diffusion in self-consistent space charge fields. This profile is determined by the ionization rate profile and the chamber geometry. Energetic metal atoms sputtered from the target (with a radially nonuniform profile) thermalize in the background gas, thus heating the gas. Heat conduction to the cooler walls introduces temperature gradients in the neutral gas, which in turn introduces neutral number density gradients. This affects the ionization rate profile and therefore the electron and ion density profiles. Thermalized metal atoms diffuse to walls where they deposit. Concentration gradients in neutral metal atom density are required to drive this diffusion flux. Neutral metal atoms are also lost to ionization, and the metal ions are transported to walls by the space charge potential profiles. This set of coupled processes will result in some radial profile in total deposition rate across the wafer and in some profile in the fraction of total deposition rate that is due to ions. The goal is to find a set of conditions that give high rates of deposition uniformly across the wafer and often with as high a fraction of metal ion component as possible. The latter requirement is necessary to enhance bottom coverage in trenches and vias.

Finally, the processes that determine the microfeature shape evolution include the composition, flux, energy, and angular distributions of ions, and of neutral metal atoms. In addition, the energy and angle-dependent ion scattering and sputtering coefficients, and the neutral metal sticking coefficients, will determine how the metal film shape evolves within a feature during the deposition process.

We describe in Section II the model equations, boundary conditions, physical data, and simulation methodology. In Section III, we present some typical results and compare them to experiment results where possible. We also discuss some processing alternatives and conclude with a feature shape evolution prediction.

II. Model Description

A. FLUID PLASMA MODEL

In the fluid model, the plasma is treated as a number of separate fluids, each of which is described in terms of the first three moments of the distribution function (number density, momentum, and energy). In the past decade, RF parallel plate discharges[2–8] magnetron discharges,[9] and electron cyclotron resonance (ECR) reactors[10] have been simulated by fluid models. Fluid–particle "hybrid" codes have also been developed for modeling RF discharges,[11] plasma–chemical vapor deposition systems,[12] and ECR reactors.[13] Recently, inductively coupled plasma reactors have been studied intensively by various hybrid schemes.[14–19] Also, a hybrid scheme has been used to study an I-PVD system.[20]

The fluid model employed here for the discharge has been described in detail elsewhere,[1,21] so only a brief description will be given here. In essence, the fluid model assumes that each charged and neutral species (electrons, positive and negative ions, and all neutrals) can be described with a velocity distribution function assuming a (separate) shifted Maxwell–Boltzmann form. The directed component of velocity for each species is obtained from species momentum balance equations, and these incorporate terms for interspecies momentum transfer. The mean thermal energy (or "temperature") for each species is derived from the solution of a species energy balance equation. For the neutral species, we generally assume that there is a single temperature characterizing these species, although this assumption can be relaxed. Finally, the species number density results from solution of a species continuity equation. Appropriate boundary conditions are developed and applied for each of the equations.

1. *Ion Equations*

The equations of continuity, momentum, and energy for each ionic species are as follows:

$$\frac{\partial n_i}{\partial t} + \nabla \cdot n_i u_i = R_i \tag{1}$$

$$\frac{\partial n_i m_i \mathbf{u}_i}{\partial t} + \nabla \cdot (n_i m_i \mathbf{u}_i \mathbf{u}_i) = -\nabla p_i + Z_i e n_i \mathbf{E} + \mathbf{M}_i \tag{2}$$

$$\frac{\partial}{\partial t}(n_i C_{v,i} T_i) + \nabla \cdot (n_i C_{v,i} T_i \mathbf{u}_i) = -\nabla \cdot \mathbf{q}_i - p_i(\nabla \cdot \mathbf{u}_i) + E_i, \tag{3}$$

where n_i is the ion density, m_i is the ion mass, $\mathbf{u}_i$ is the ion velocity, T_i is the ion temperature, p_i is the ion pressure (static pressure), Z_i is the ion charge, and $\mathbf{E}$ is the electrostatic field. $C_{v,i}$ is the ion heat capacity at constant volume. The ion heat flux vector, $\mathbf{q}_i$, is assumed to be given by

$$\mathbf{q}_i = -\frac{5}{2}\frac{n_i k T_i}{m_i \nu_{in}^m}\nabla(kT_i), \tag{4}$$

where ν_{in}^m is the ion collison frequency with neutrals. The terms R_i, $\mathbf{M}_i$, and E_i represent the transfer of mass, momentum, and energy, respectively, to the ions by collisions with neutral species.

The boundary conditions for the previous equations are as follows. At the centerline of the simulation all gradients are zero by symmetry. At the walls, the gradient of positive ion flux normal to the wall is set to zero. From this, the velocity is found by linear extrapolation, and the density is determined from this value. Since the negative ions are massive and relatively cold, they are trapped by the electrostatic potential; therefore, the negative ion flux, velocity, and density are set to zero at walls. Finally, all ion temperature gradients are set to zero at walls (no thermal conduction).

2. *Thermalized Neutral Species Equations*

By neglecting convective transport, the thermalized neutral species density can be obtained by solving the diffusion equations for each species:

$$\frac{\partial n_{\mathrm{n}}}{\partial t} + \nabla \cdot \mathbf{\Gamma}_{\mathrm{n}} = R_{\mathrm{n}}, \tag{5}$$

where $\mathbf{\Gamma}_{\mathrm{n}}$ is the neutral flux driven by the number density gradient ∇n_{n} and temperature gradient ∇T:

$$\mathbf{\Gamma}_{\mathrm{n}} = n_{\mathrm{n}}\mathbf{u}_{\mathrm{n}} = n_{\mathrm{n}}\left[-D_{\mathrm{n}}\frac{\nabla p_n}{p_n}\right] = -D_{\mathrm{n}}\left[\nabla n_{\mathrm{n}} + n_{\mathrm{n}}\frac{\nabla T}{T}\right], \tag{6}$$

where n_{n} is the neutral density, m_{n} is the neutral mass, $\mathbf{u}_{\mathrm{n}}$ is the neutral species velocity, p_n is the partial pressure of the nth neutral species, and D_{n} is the diffusion coefficient. By assuming that the neutral species are in local thermal equilibrium characterized by a single temperature, a single energy balance equation for the neutral gas is written as

$$\frac{\partial}{\partial t}(nC_{\mathrm{v}}T) = -\nabla \cdot \mathbf{q} + \sum_n E_{\mathrm{n}}, \tag{7}$$

where $n = \Sigma_n n_{\mathrm{n}}$ is the total thermalized neutral density, and T is the temperature, $C_{\mathrm{v}} = \Sigma_n (C_{\mathrm{v}})_n (n_{\mathrm{n}}/n)$ is the average heat capacity. The total neutral heat flux vector, $\mathbf{q} = \Sigma_n \mathbf{q}_{\mathrm{n}}$, is defined analogously to that for ions (Eq. 4). All neutral species are assumed to be in their ground electronic state. The terms R_{n} and E_{n} represent the transfer of mass and energy, respectively, to the slow thermalized species.

The neutral flux at the wall for species n (except metal species) is the sum of several sources:

$$\Gamma_{\mathrm{n}} = n_{\mathrm{n}}\mathbf{u}_{\mathrm{n}} \cdot \hat{n} = \Gamma_{\mathrm{n,c}} + \Gamma_{\mathrm{n,i}} + \Gamma_{\mathrm{F}} - \Gamma_{\mathrm{P}}, \tag{8}$$

where $\Gamma_{\mathrm{n,c}}$ is the flux from chemical reactions on the walls, $\Gamma_{\mathrm{n,i}}$ is the ion recombination at walls, Γ_{F} is the gas feed, and Γ_{P} is the gas pumping. The latter two terms are nonzero only for inflow or outflow wall locations, respectively. For all species except the metal species, we assume that the various ions are neutralized at the wall to form the parent neutral. The gas pumping flux is calculated for the wall area designated as the pump port by

$$\Gamma_P = \tfrac{1}{4} n_n \bar{v}_n (1 - R), \tag{9}$$

where R is the fraction of incident molecules reflected at the port and is adjusted in the simulation to maintain a specified total neutral pressure. $\bar{v}_n = \sqrt{8kT/\pi m_n}$ is the mean thermal speed for species n.

For metal species, due to the assumed unity sticking coefficient on the wall, the flux of metal atoms to the reactor wall is $\Gamma_n = -\frac{1}{4} n_n \bar{v}_n$. All chamber walls act like a pump port without reflection, i.e., $R = 0$. This is a strong sink for metal atoms. We also assume metal ions deposit with unity probability at the wall.

3. *Electron Equations*

The behavior of the electrons is governed by the equations of continuity and energy:

$$\frac{\partial n_e}{\partial t} + \nabla \cdot \mathbf{\Gamma}_e = R_e \tag{10}$$

$$\frac{\partial}{\partial t}\left(\frac{3}{2} n_e k T_e\right) = -\nabla \cdot \mathbf{Q}_e - e\mathbf{E} \cdot \mathbf{\Gamma}_e + P_{\text{ind}} + E_e, \tag{11}$$

where

$$\mathbf{\Gamma}_e = -\frac{1}{m_e \nu_{\text{en}}^{\text{m}}} \nabla p_e - \frac{\text{e}n_e}{m_e \nu_{\text{en}}^{\text{m}}} \mathbf{E} \tag{12}$$

is the electron flux by the drift-diffusion form of the electron momentum balance, and

$$\mathbf{Q}_e = \frac{5}{2} \mathbf{\Gamma}_e k T_e - \frac{5}{2} \frac{n_e k T_e}{m_e \nu_{\text{en}}^{\text{m}}} \nabla (k T_e) \tag{13}$$

is the electron energy flux assuming negligible directed energy. n_e is the electron density, m_e is the electron mass, T_e is the electron temperature, p_e is the electron pressure, and $\nu_{\text{en}}^{\text{m}}$ is the electron momentum transfer collision frequency with neutrals. The terms R_e and E_e represent the transfer of mass and energy to the electrons by collisions with other species. P_{ind} is the RF period-averaged inductive heating term:

$$P_{\text{ind}} = \tfrac{1}{2} \text{Re}(\sigma_{\text{p}} |\varepsilon|^2), \tag{14}$$

where σ_{p} is the plasma conductivity and ε is the electric field intensity.

The boundary conditions for Eqs. (10) and (11) are

$$\mathbf{\Gamma}_e = \tfrac{1}{4} n_e \bar{v}_e \exp(\phi_s / k T_e) \tag{15}$$

and

$$\mathbf{Q}_e = 2\mathbf{\Gamma}_e kT_e, \tag{16}$$

where $\bar{v}_e$ is the electron average speed and ϕ_s is the wall sheath potential.

4. Electromagnetic Equations

We treat separately the steady-state electric field due to space charge within the plasma and the time-varying electric field from the inductive power coupling. The former is determined by solving Poisson's equation:

$$\varepsilon_0 \nabla^2 \phi = e(n_e - \sum_i Z_i n_i), \tag{17}$$

where ε_0 is the permittivity of free space and ϕ is the electrostatic potential.

Based on the assumption of azimuthal symmetry of the electromagnetic field, Maxwell's equations reduce to Laplace's equation for the radial and axial electric fields and the Helmholtz wave equation for the azimuthal electric field, ε_θ, with the external current $J_{\text{ext},\theta}$ as a forcing function

$$\nabla^2 \varepsilon_\theta + \frac{\omega^2}{c^2} K \varepsilon_\theta = -i\omega\mu_0 J_{\text{ext},\theta}, \tag{18}$$

where μ_0 is the permeability of free space.

For constant power operation, the external coil currents are iterated upon until the volume integrated inductive power calculated by Eq. (14) meets that specified for the simulation. The coil current may also be kept constant, with the power deposited being dependent on plasma conditions. In principle, it is also possible to compute capacitive coupling between coils and the plasma, but we have not included these calculations.

Our model assumes that all power is deposited collisionally. Collisionless heating may be an important mechanism, especially at low pressures[22] or high RF frequency.[23] The major effect of neglecting collisionless heating is to improperly predict the magnitude of ε_θ. However, in the current model, we generally adjust the external coil current in order to achieve a target total inductive power deposited into electrons. At a given value of total deposited power, the solution is insensitive to the value of ε_θ required to achieve that power.

5. Mass, Momentum, and Energy Transfer in Fluid Model

The mass balance collision term gives the rates of creation and loss of a species. For the formation of species α by an electron collision with neutral

β, we use the expression

$$R_\alpha = \sum_r l_{r,\alpha} k_r(T_e) n_e n_\beta, \tag{19}$$

where $l_{r,\alpha}$ is the number of particles of species α created or lost per collision of type r,

$$k_r(T_e) = \int \sigma_r(v_e) v_e f_e(\mathbf{v}_e)\, d^3 v_e \tag{20}$$

is the reaction rate coefficient, $\sigma_r(v_e)$ is the collision cross section for the particular reaction, and $f_e(\mathbf{v}_e)$ is the electron velocity distribution, assumed to be Maxwellian in the fluid model. The integration is performed using cross sections from the literature to generate reaction rate coefficients as a function of electron temperature. These are listed in Tables 1 and 2.

For metal atoms, due to the thermalization of the sputtered metal atoms, an additional source term $R_{\text{therm}}^{\text{metal}}$ is added to the thermalized neutral species continuity equation. This is discussed in Section II,C.

The electron-neutral momentum transfer collision frequency used in Eqs. (12) and (13) is given by

$$\nu_{\text{en}}^{\text{m}} = \sum n_\beta \int f_e(\mathbf{v}_e) v_e \sigma_{e\beta}^{\text{m}}(v_e)\, d^3 v_e, \tag{21}$$

where the summation is over all neutrals that have collisions with the electrons, and the momentum transfer cross section, $\sigma_{e\beta}^{\text{m}}(v_e)$, is obtained from the literature (see Tables 1 and 2). We have assumed that the use of the period-averaged electron energy gives collision rates that are not too different from those which would be obtained by resolving the electron energy variation over the RF period of the inductive coil current.

TABLE 1
ELECTRON Ar COLLISION DATA

Reaction	Type	ε_r(eV)	Reference
$e + Ar \rightarrow Ar^+ + e + e$	Ionization	15.60	Peterson and Allen[24]
$e + Ar \rightarrow Ar + e$	Momentum transfer	$\frac{3\, m_e}{m_n} T_e$	Peterson and Allen[24]
$e + Ar \rightarrow Ar^* + e$	Electronic excitation	11.60	Peterson and Allen[24]
$e + Ar \rightarrow Ar^* + e$	Electronic excitation	14.25	Peterson and Allen[24]
$e + Ar \rightarrow Ar^* + e$	Electronic excitation	14.79	Peterson and Allen[24]
$e + Ar \rightarrow Ar^* + e$	Electronic excitation	15.48	Peterson and Allen[24]

TABLE 2
ELECTRON Al COLLISION DATA

Reaction	Type	ε_r(eV)	Reference
$e + \mathrm{Al} \rightarrow \mathrm{Al} + e$	Momentum transfer (same as Ar)	$\frac{3m_e}{m_n} T_e$	Peterson and Allen[24]
$e + \mathrm{Al} \rightarrow \mathrm{Al}^+ + e + e$	Ionization	5.97	Shimon *et al.*[25]
$e + \mathrm{Al} \rightarrow \mathrm{Al} + e$	Electronic excitation	3.00	Dickson *et al.*[26]

The collisional transfer of momentum from ion species α to neutral species β is handled explicitly through the term $\mathbf{M}_i$. These terms are derived by integrating the momentum transfer cross section $\sigma_{\alpha\beta}^{m}$ over the assumed drifting Maxwellian ion and neutral velocity distribution functions:

$$\mathbf{M}_i = \sum_{\beta} -\frac{m_i m_\beta}{m_i + m_\beta} n_i \mathbf{u}_i \nu_{i\beta}^{m}, \tag{22}$$

where

$$\nu_{i\beta}^{m} = n_\beta \sigma_{i\beta}^{m} \left(\frac{16}{9} \frac{8kT_i}{\pi m_i} + u_i^2 \right)^{1/2}, \tag{23}$$

and the summations are carried out over all ions α or all neutrals β. The momentum transfer cross sections $\sigma_{\alpha\beta}^{m}$ are taken from the literature or estimated from the hard sphere radii of the collision partners. Generally, the ion-neutral momentum transfer cross sections are functions of the ion and neutral temperatures and velocities. However, they are currently approximated as constant.

For the electrons, we calculate the total rate of energy loss from collisions as

$$E_e = \sum_r \varepsilon_r k_r(T_e) n_e n_\beta, \tag{24}$$

where ε_r is the average energy lost from the electrons per collision of type r. The value of ε_r for each reaction is listed in Tables 1 and 2.

The ion and neutral collisional energy loss terms are

$$E_\alpha = \frac{1}{2} m_\alpha u_\alpha^2 R_\alpha - \mathbf{M}_\alpha \cdot \mathbf{u}_\alpha + \sum_\beta \left. \frac{\delta(n_\alpha \langle K_\alpha \rangle)}{\delta t} \right|_\beta . \tag{25}$$

The first and second terms arise from the transformation of the moment equation for total (thermal and directed) energy into an equation for thermal energy. The third term is the collision integral for the transfer of total energy, $\langle K_\alpha \rangle$, to species α from species β, given by

$$\left.\frac{\delta n_n \langle K_{\rm n} \rangle}{\delta t}\right|_i = -\left.\frac{\delta n_{\rm i} \langle K_{\rm i} \rangle}{\delta t}\right|_n = n_{\rm i} n_{\rm n} \sigma_{\rm in}^{\rm m} \frac{2 m_{\rm i} m_{\rm n}}{(m_{\rm i} + m_{\rm n})^2} \times \left\{ \frac{m_{\rm i}}{2} \left(\frac{\pi^{2/3}}{2^{2/3}} \frac{8kT_{\rm i}}{\pi m_{\rm i}} + u_{\rm i}^2 \right)^{3/2} - \left(\frac{8kT_{\rm i}}{\pi m_{\rm i}} + u_{\rm i}^2 \right)^{1/2} \frac{3}{2} kT_{\rm n} \right\}. \quad (26)$$

$E_{\rm n}$ also includes heating of the neutral fluid by chemical reactions and the heating of fast neutrals by collisions (energy transfer during the thermalization process of fast neutrals $E_{\rm en}^{\rm fast}$) which are discussed later in this section. In addition, the neutral species may be heated by such effects as neutralized ions energetically reflecting from walls, but this effect has been neglected in the current model.

6. *Limitations of the Fluid Model*

There are several parts of I-PVD plasma that are not well described by the fluid equations. Recall that the basic assumption of the fluid model is that the velocity distribution function for each species could be properly represented using a Maxwell–Boltzmann function, with the velocity shifted by a mean velocity term. This is sometimes referred to as a "shifted Maxwellian" distribution. It is well-known that electrons do not always follow this distribution, although in some cases the Maxwellian distribution is a reasonable approximation. We have developed a scheme to couple the inhomogeneous Boltzmann equation solution for electrons to the fluid model, solving for the electron energy distribution function (EEDF) iteratively with the fluid model which is described elsewhere.[19] In the results presented here, we retain the assumption that the electrons follow the Maxwell–Boltzmann distribution. In addition to electrons, metal atoms that are sputtered from the target are not well described by a shifted Maxwellian distribution, and for these species we have chosen to implement a fast-particle Monte Carlo simulation, which is described in Section II,B. When fast metal atoms have thermalized sufficiently through collisions with slower species, the simulation converts "fast" metal atoms into "thermalized" metal atoms. The latter species are described using Eqs. (5)–(9) given in Section II,A,2. The hybrid strategy for coupling the fluid plasma model and fast-particle Monte Carlo simulation is described in Section II,C. Finally, at the end of the plasma simulation, when we use the plasma model results as input into a feature profile simulation, an ion Monte Carlo simulation is performed (Section II,D). The results of this simulation provide the ion velocity distribution across the wafer radius. Although the fluid model assumptions generally work well for ions when the aim is to predict the ion density and flux profile within the plasma, the assumption of a shifted Maxwellian velocity distribution is not sufficiently accurate for a feature profile simulation. The ion Monte Carlo simulation may be coupled to an RF sheath model if RF bias is used on the substrate.

B. FAST NEUTRAL THERMALIZATION: MONTE CARLO SIMULATIONS

Under the bombardment of the high-energy plasma species (most are positive ions), the metal atoms sputtered from the target (or coils) generally follow the Thompson energy distribution and cosine angular distribution $f(E, \theta)$:[27]

$$f(E, \theta) = \frac{2UE}{(U + E)^3} \frac{\cos \theta}{\pi}, \tag{27}$$

where U is the surface bonding energy of the target material, usually taken to be equal to the sublimation energy (e.g., 3.38 eV for aluminum),[28] and θ is the angle measured from the surface normal. Thus, the energy of the sputtered atoms could range from room temperature to as high as several tens of electron volts, and a significant fraction of sputtered atoms energies are on the order of several electron volts.

The sputtered metal atoms will move through the plasma until they are ionized or reach the substrate or some other surface. Generally, under relatively low discharge pressure conditions (<0.1 mTorr), they move ballistically in the plasma after leaving the target, with few collisions with discharge gas atoms. At higher discharge pressures (>50 mTorr), on the other hand, sputtered atoms are thermalized rapidly due to the frequent collisions with the background gas species. They then travel through the chamber primarily by diffusion until they reach a wall where they deposit. There exists an intermediate pressure range (0.5–30 mTorr), though, in which sputtered atoms are considerably influenced by energy and momentum exchange with background gas atoms. Typically, the I-PVD system runs in this intermediate pressure regime.

Considering the interaction between the sputtered atoms and the background discharge gas atoms, a considerable amount of energy and momentum is imparted to the background neutral gas atoms during the collisions. As a consequence, the fast sputtered atoms lose their energy and slow down (i.e., thermalize), and simultaneously the background neutrals are heated. This effect leads to an enhanced gas dynamic flow, referred to as a "sputter wind," and a reduction in the neutral gas density in the vicinity of the target. A sputter wind has been measured by Hoffman,[29] and the gas density reduction in front of the magnetron target has been measured by Rossnagel.[30]

The transport behavior of sputtered atoms can be conveniently simulated by the fast-particle Monte Carlo method. The fast-particle Monte Carlo method has been extensively used in the simulation of various sputtering systems.[28,31–36] The particle Monte Carlo method is carried out in 3-D (x, y, and z coordinates) and (3-V) (v_x, v_y, and v_z).

Details of the fast-particle Monte Carlo method used here can be obtained elsewhere.[31,37–40] Briefly, when a sputtered atom leaves the target, the energy and angle of departure are determined by the selection of two random numbers. Collisions are treated using the "mean free path method." Scattering angle and energy are determined from a distribution assuming a power law interatomic potential and isotropic scattering in the center-of-mass frame.

The atom is absorbed if the atom hits a wall. Otherwise, the atom is tracked until its energy decreases to the same thermal energy as the local background gas, i.e., $\frac{3}{2}kT(r, z)$, where $T(r, z)$ is the neutral temperature obtained from the energy conservation equation (Eq. 7). At this point, the atom joins the thermalized fluid group.

In each sputtering cell of the target, 10,000 superparticles are released into the plasma. Due to the different areas of the sputtering cells, a weighting function corresponding to the area of sputtering cells is defined by $W_j = (\Gamma_j^{\text{spu}} \cdot S_j)/10{,}000$ (with unit of atoms/sec) for the calculation of the thermalization rate profile, where S_j is the sputtering area of the jth sputtering cell, and Γ_j^{spu} is the sputtered atom flux at cell j.

C. Hybrid Strategy for Neutral Transport

The "hybrid model" of neutral transport is illustrated in Fig. 1 and can be summarized as follows: (i) Make an initial guess of the plasma profiles; (ii) conduct the fast neutral Monte Carlo simulation based on the slow neutral density profiles n_{neu}, temperature profile T, electron density profile n_e, and electron temperature T_e. Fast neutral thermalization rate ($R_{\text{therm}}^{\text{fast}}$), the energy transfer from fast neutrals to the slow background gases ($R_{\text{en}}^{\text{fast}}$), and the ionization of fast neutrals by electron impact ($R_{\text{ion}}^{\text{fast}}$) are determined; (iii) couple $R_{\text{therm}}^{\text{fast}}$ and $R_{\text{en}}^{\text{fast}}$ into the slow neutral fluid equations and $R_{\text{ion}}^{\text{fast}}$ into the electron and ion continuity equations. Recompute the plasma and fluid equations to obtain new profiles of n_{neu}, T, n_e, and T_e. Steps (ii) and (iii) are repeated until the results converge to a steady solution. Note that it is not necessary to conduct the Monte Carlo simulation for each time step because the thermalization of fast neutrals is mostly determined by the slow

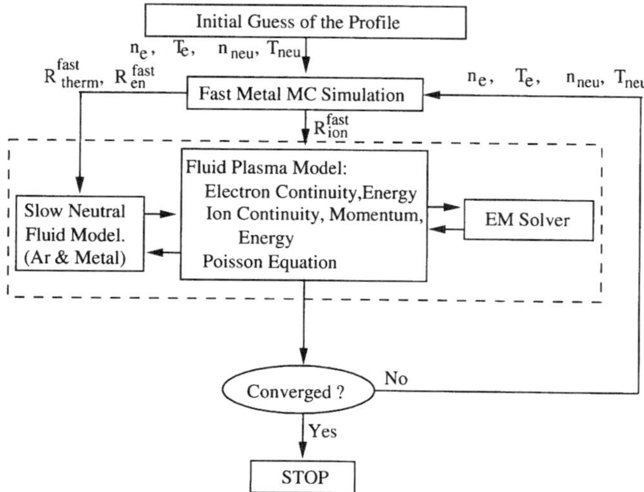

FIG. 1. Scheme of the hybrid model of neutrals transport.

background neutrals. Thus, the fast neutral profiles change only when the slow neutral profile changes significantly. Generally, the Monte Carlo simulation is run after running the plasma model for 5 μsec.

D. POSTPROCESSING MONTE CARLO FOR NEUTRAL AND IONIC ANGULAR DISTRIBUTION FUNCTIONS AT SUBSTRATE

After the simulation reaches convergence, an ion Monte Carlo simulation is used in the bulk plasma to obtain the ion energy distribution (IED) and ion angular distribution (IAD) of metal ions at the sheath edge. Then an analytical sheath model[41] is used to calculate the IED and IAD on the substrate after the ions cross the thin sheath region.

The ion Monte Carlo method we employ is similar to the one described for fast neutrals. Elastic and charge exchange ion-neutral collisions are treated. Particle ions are launched throughout the bulk plasma at a rate proportional to the local rate of ionization. Ions are accelerated by the space charge electric field and scatter after collisions with neutrals. We note that the ion Monte Carlo simulation does not affect any other part of the plasma simulation and is conducted after the simulation has reached steady state. The purpose is to provide IED and IAD for metal ions to the profile simulator.

E. Feature Profile Evolution Model

The feature profile evolution simulation we have developed is a string-type model in which we discretize a 2-D cross section of a trench or via with a series of nodes. While many researchers have developed such models, our work is based largely on the work of Dalvie *et al.*[42] These authors show that a 2-D/3-V (two space dimensions and three velocity dimensions) treatment of the particle flux integration is necessary for the proper calculation of geometric shadowing of incident ions and neutrals by feature walls. Additionally and more subtly, a 2-D/3-V approach is necessary in order to differentiate between trench and via evolution. Several authors[43,44] have used a 2-D/2-V treatment to study feature profile evolution during both etching and deposition; however, this is aphysical since it implies that the incident fluxes are constrained to the two dimensions given by the feature geometry. No allowance is made for proper shadowing in the third, symmetric, dimension, e.g., the curved walls of a via or the "infinitely" long walls of a trench. Since in this work we are concerned with details of trench filling by metal ions as well as thermal and directed neutrals (which depend sensitively on the plasma-based fluxes of these species), a rigorous 2-D/3-V approach was chosen in order to ensure a more accurate flux integration.

1. Plasma Flux Integration

A schematic diagram of the trench geometry we employ is shown in Fig. 2. The y direction points into the plasma, with the x–y plane being the symmetry plane. If $f(v, \theta, \phi)$ is the distribution function in spherical coordi-

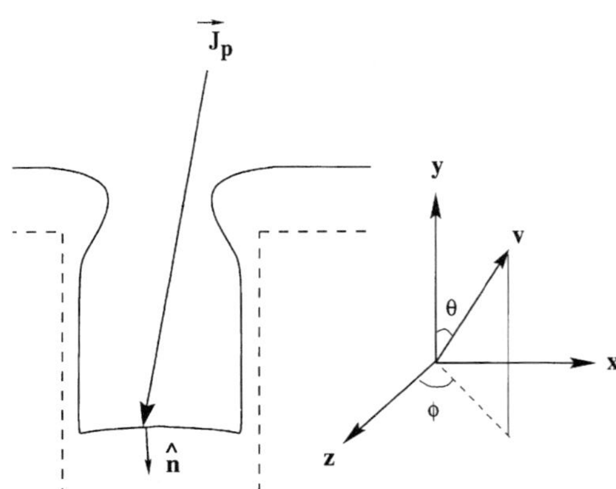

FIG. 2. Schematic diagram of the trench geometry and coordinate system.

nates for a given ionic or neutral species, then at a reference x–z plane arbitrarily close to the mask top shown in Fig. 2, we have in general for the x and y components of the particle flux crossing that plane

$$J_{\mathrm{p},x} = \int_0^{2\pi} \int_0^{\pi/2} \int_0^{\infty} nv \sin\theta f(v, \theta, \phi) v^2 \sin\theta \, dv \, d\theta \, d\phi \tag{28}$$

and

$$J_{\mathrm{p},y} = \int_0^{2\pi} \int_0^{\pi/2} \int_0^{\infty} nv \cos\theta f(v, \theta, \phi) v^2 \sin\theta \, dv \, d\theta \, d\phi. \tag{29}$$

In a trench or via geometry in which axisymmetry can be assumed, f is not a function of ϕ. Taking this into account, we define a function $F_{\mathrm{p}}(\theta)$ as

$$F_{\mathrm{p}}(\theta) = 2\pi \sin\theta \int_0^{\infty} nv \cos\theta f(v, \theta) v^2 \, dv. \tag{30}$$

We can then write

$$J_{\mathrm{p},y} = -\int_0^{\pi/2} F_{\mathrm{p}}(\theta) \, d\theta. \tag{31}$$

This shows that $F_{\mathrm{p}}(\theta)\, d\theta$ is the differential flux passing between θ and $\theta + d\theta$ and is therefore the function one obtains when performing an angular binning procedure in the ion Monte Carlo code discussed in Section II,D. For an arbitrary shadowed surface in which the ϕ integration limits are functions both of surface position $\mathbf{r}$ and of polar angle θ,[42] one obtains, after substituting Eq. (30), the general axisymmetric versions of Eqs. (28) and (29):

$$J_{\mathrm{p},x}(\mathbf{r}) = \frac{1}{2\pi} \int_{\phi_{\mathrm{l}}(\mathbf{r})}^{\phi_{\mathrm{u}}(\mathbf{r})} \int_0^{\pi/2} \cos\phi \tan\theta F_{\mathrm{p}}(\theta) \, d\theta \, d\phi \tag{32}$$

and

$$J_{\mathrm{p},y}(\mathbf{r}) = \frac{1}{2\pi} \int_{\phi_{\mathrm{l}}(\mathbf{r})}^{\phi_{\mathrm{u}}(\mathbf{r})} \int_0^{\pi/2} F_{\mathrm{p}}(\theta) \, d\theta \, d\phi. \tag{33}$$

As shown by Dalvie *et al.*,[42] analytic expressions for $\phi_{\mathrm{l}}(\mathbf{r})$ and $\phi_{\mathrm{u}}(\mathbf{r})$ as functions of θ for both trenches and vias are available. Thus, the ϕ integral in each equation can be evaluated analytically, leaving integrations only over θ. For both ions and neutrals, Eqs. (32) and (33) are numerically integrated using Simpson's rule in order to obtain the flux components impinging at each surface position $\mathbf{r}$. For ions and fast neutrals, $F_{\mathrm{p}}(\theta)$ isobtained by binning over θ in the ion Monte Carlo code as discussed previously. For thermalized neutrals, the flux distribution is assumed to be isotropic (Maxwellian), and therefore

$$F_p(\theta) = n\left(\frac{2kT}{\pi m}\right)^{1/2} \sin\theta\cos\theta, \tag{34}$$

where m is the neutral mass and T is the neutral gas temperature.

2. *Chemistry Model*

Once number fluxes for all impacting species are known as a function of position along the evolving surface, the deposition rate at each advancing node must be calculated. We assume that the deposition rate at surface position $\mathbf{r}$ is given by

$$\mathrm{DR}(\mathbf{r}) = \{\mathbf{J}_{\mathrm{Al}^+}(\mathbf{r})\cdot\hat{n}(\mathbf{r}) + \mathbf{J}_{\mathrm{Al}_f}(\mathbf{r})\cdot\hat{n}(\mathbf{r}) + \mathbf{J}_{\mathrm{Al}_t}(\mathbf{r})\cdot\hat{n}(\mathbf{r})\}/\rho_{\mathrm{Al}}, \tag{35}$$

where $\mathbf{J}_i(\mathbf{r})$ are the fluxes of i at $\mathbf{r}$, and ρ_{Al} is the density of aluminum. For the current study we assume a unity sticking coefficient for all species regardless of incident energy or angle, and sputtering is neglected.

Equation (35) gives the deposition rate at the segments since we know ion and neutral number fluxes at the segments. The nodal deposition rates are calculated using the angle bisector method. That is, the nodal deposition rate at surface position $\mathbf{r}$, $\mathrm{DR}_{\mathrm{n}}(\mathbf{r})$, is taken to be the average of the two adjacent segment deposition rates, and the nodes are moved along their unit normal $\hat{n}_{\mathrm{n}}(\mathbf{r}) = [n_{\mathrm{n},x}(\mathbf{r}), n_{\mathrm{n},y}(\mathbf{r})]$ using the following expressions:

$$d_x(\mathbf{r}) = \mathrm{DR}_{\mathrm{n}}(\mathbf{r}) \times n_{n,x}(\mathbf{r}) \tag{36}$$

and

$$d_y(\mathbf{r}) = \mathrm{DR}_{\mathrm{n}}(\mathbf{r}) \times n_{\mathrm{n},y}(\mathbf{r}), \tag{37}$$

where $d_x(\mathbf{r})$ and $d_y(\mathbf{r})$ are the nodal displacements in x and y at $\mathbf{r}$.

III. Simulation Results for Aluminum Film Deposition

A. SIMULATION CONDITIONS

In order to investigate the physics of I-PVD systems, a relatively chemically simple system, namely, a single-component metal film in an inert gas environment, has been chosen for study. The simulation results shown here are for aluminum film deposition with argon. The I-PVD system is excited with a two-turn RF coil configuration, as shown in Fig. 3. The frequency of

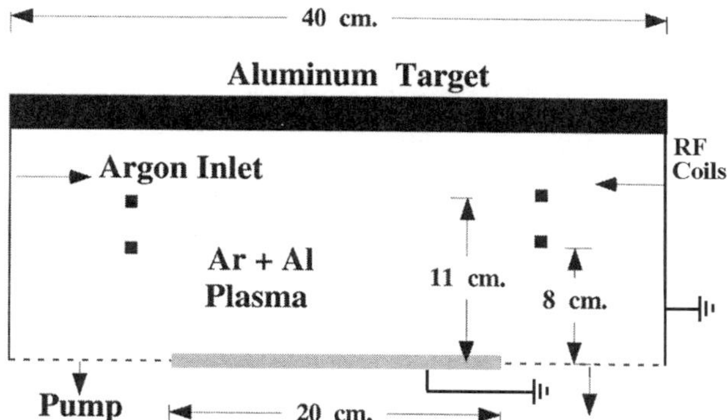

FIG. 3. Sectional view of an I-PVD reactor geometry for pure metal (aluminum) film deposition in argon and aluminum inductively coupled plasma system.

the RF current flowing through the coil is 13.56 MHz. No capacitive coupling from the coil is considered. The chamber is cylindrical and axisymmetric with a 20-cm radius (r) and 15-cm height (z). The 20-cm substrate is located 15 cm from the target on the top. We assume all the walls are conducting and grounded, and the wall temperature is assumed to be room temperature. Generally, the bias applied to the substrate is low (~ 10 V), and the effects on the bulk plasma are negligible. The argon is introduced into the reactor from the side wall with a constant flow rate of 100 sccm. A uniform aluminum atom flux 10^{16} cm^2 sec (corresponding to an equivalent sputtering "flow rate" of 30.61 sccm) is sputtered from the target. The total power deposition into the discharge is set at 500, 1000, and 1500 W, and the total neutral pressure is kept at 35 mTorr.

B. TYPICAL RESULTS AND BASIC CHARACTERISTICS OF I-PVD PLASMAS

A typical set of results of the I-PVD plasma is shown in Fig. 4 for the configuration described previously. In Fig. 4, the total RF power deposited in the plasma is 1000 W. The electron density n_e, electron temperature T_e, plasma potential Φ, and power deposition profiles P_{rf} are very similar to those of a pure argon plasma due to the relatively low concentration of metal species. The average percentage aluminum concentration is 0.04% in this case. It is not surprising that most of the RF power is deposited in a small region in which the coil is located because the inductive RF current in the plasma is confined to a thin skin layer. The plasma potential profile is flat in the bulk plasma and peaks at about 17.3 V. The peak electron density is 7×10^{11} cm^3 and is located on the chamber axis. The electron

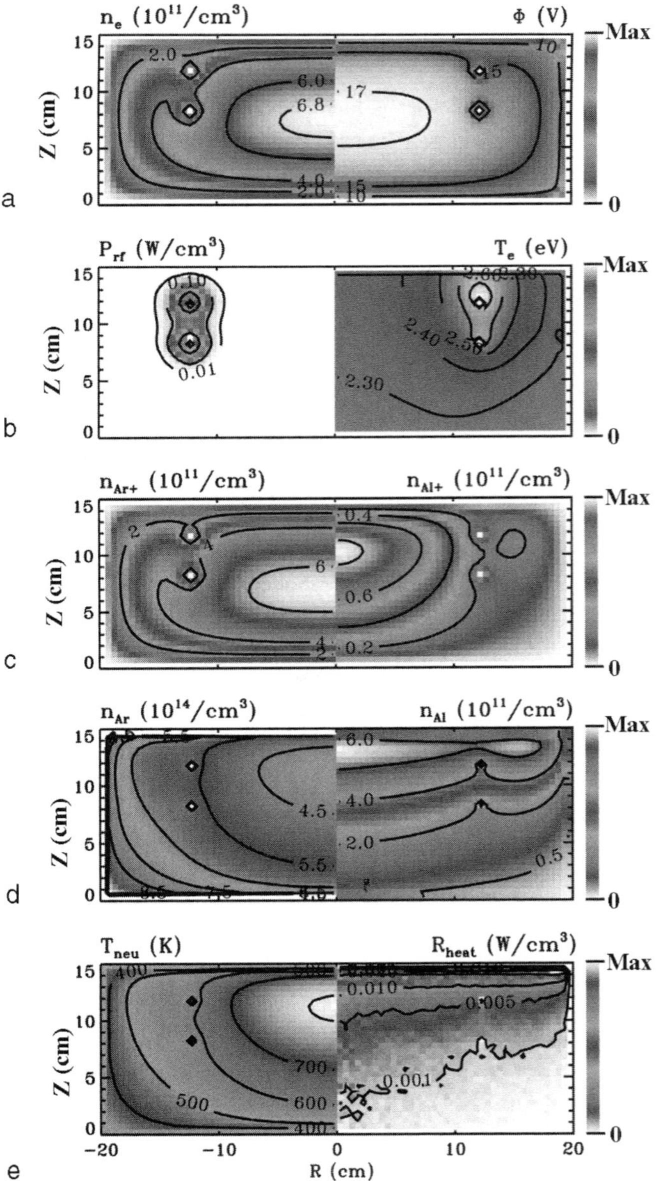

FIG. 4. Profiles of (a) electron density n_e and plasma potential Φ, (b) RF power deposition P_{rf} and electron temperature T_e, (c) ion densities n_{Ar^+} and n_{Al^+}, (d) neutral densities n_{Ar} and n_{Al}, and (e) neutral temperature T_{neu} and the energy transfer rate R_{heat} from fast atoms to slow background gases.

temperature ($\sim$2.5 eV) across the chamber is relatively flat, with only a slight increase near the RF coils.

The metal neutral and ion species density profiles are among the most important variables in I-PVD systems. The metal species flux (atoms and ions) will determine the film deposition rate across the wafer as well as the deposition profile within features. However, it should be noted that the fast metal atom thermalization and the subsequent slow metal atom transport behavior are greatly affected by the background neutral species density and temperature. As mentioned in Section II, due to the heat transfer from fast-sputtered atoms to the background gas, a “gas rarefaction” or “sputtering wind” effect is known to occur in sputtering. The fast neutral species Monte Carlo simulations show a much stronger heating near the target than in the downstream region around the substrate. As shown in Fig. 4e, the energy transfer rate to the background atoms from the fast sputtered atoms, R_{heat}, in front of the target, is more than one order of magnitude larger than that near the substrate. The neutral gas temperature near the target can be as high as 800 K. Simultaneously, due to the large diffusion coefficient of neutral species under low-pressure conditions, the pressure is nearly constant across the chamber. As a result, the neutral density is lower in the high-temperature region since the ideal gas law states $n = p/kT$. The Ar and Al neutral atom densities are both affected by this gas rarefaction effect. A large fraction of fast sputtered Al atoms are thermalized near the target and a peak in the density of the thermalizing Al atoms is found about 1 or 2 cm below the target. Both Ar^+ and Al^+ densities peak on the chamber axis. However, Al^+ peaks closer to the target due to the peak in Al density there. Furthermore, the Ar^+ density is about 10 times larger than Al^+ density due to the small concentration of Al atoms in the plasma, despite the relatively low threshold energy for Al ionization compared to Ar.

The aluminum film deposition rate and the fraction of aluminum ion flux in the total deposition flux are shown in Figs. 5a and 5b. Figure 5a plots the radial profiles of the contribution of each of the three Al deposition species: Al^+, fast Al, and slow Al. The total deposition rate is also plotted. Obviously, the film deposition rate is radially nonuniform. The film deposition rate is highest in the center of the substrate. This is mainly caused by the Al^+ flux profile. It also demonstrates that at the substrate there are unthermalized fast neutral Al atoms (about 5% of the total deposition flux), even though these species have traveled a relatively long distance from the target. It can be expected under lower pressure conditions with lower background neutral density or/and with a shorter distance from the target to substrate that the contribution of unthermalized atoms will be larger. The ratio of aluminum ion flux to the total deposition flux shown in Fig. 5b is about 0.7 and this factor is relatively uniform across the wafer.

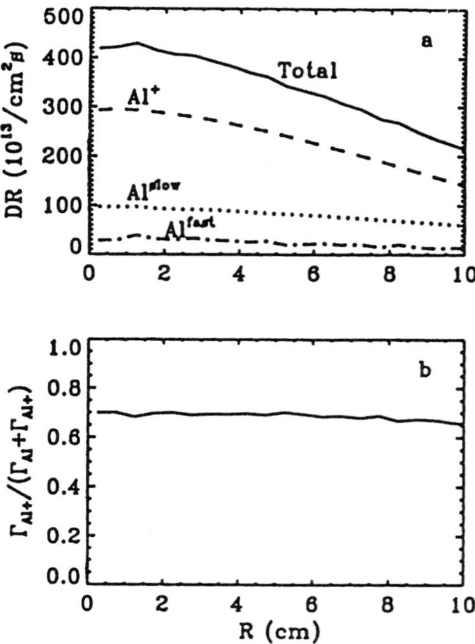

FIG. 5. (a) Total aluminum film deposition rate and contributions from each component of metal species: aluminum ions (Al^+), thermalized aluminum atoms (Al^{slow}), and unthermalized aluminum atoms (Al^{fast}). (b) Flux ratio of aluminum ions (Al^+) to the total deposition flux across the substrate.

It is helpful in understanding the key couplings in the I-PVD plasma to examine profiles along the axial centerline of the system. The 1-D profiles along the chamber axis are plotted in Figs. 6–8, respectively. Figure 6a demonstrates that the plasma potential on the chamber axis is relatively flat in the bulk plasma and it is almost symmetric about the middle point at $z = 7.5$ cm. The ions created in the body of the plasma will flow to one of the two surfaces. Ions created below $z \cong 7.5$ cm will flow to the substrate, and ions created above $z \cong 7.5$ cm will flow to the target. This implies that metal atoms ionize before they diffuse past $z \cong 7.5$ cm will return to the target in the form of metal ions.

Due to the thermalization of fast sputtered aluminum near the target, the thermalized aluminum atom density reaches its peak several centimeters from the target, and then it decreases rapidly to the wafer. Close to the wafer surface, the thermalized aluminum atom density is one order of magnitude lower than that near the target. Simultaneously, due to the almost constant electron temperature along the chamber axis shown in Fig. 7a, the Al

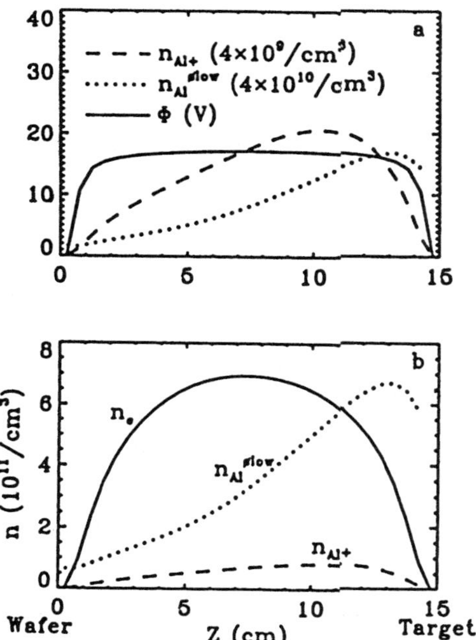

FIG. 6. 1-D profiles on chamber axis for (a) plasma potential Φ, aluminum ion density n_{Al^+}, and thermalized aluminum atom density $n_{\mathrm{Al}}^{\mathrm{slow}}$; (b) electron density n_e, aluminum ion density n_{Al^+}, and thermalized aluminum atom density $n_{\mathrm{Al}}^{\mathrm{slow}}$.

ionization rate is largest in front of the target, and thus the Al ion density also peaks near the target. This can also be seen from the 2-D profile of Al^+ density shown in Fig. 4. It is worth mentioning that about 10–15% of the aluminum is ionized in the discharge (as shown in Fig. 6b). By contrast, typically less than about 1% of the argon is ionized in inductively coupled discharges.

The negative value of metal species fluxes shown in Fig. 7b means that the direction of flux is downward to the wafer, whereas the positive value indicates the flux direction is toward the target. It is easy to understand that the plasma potential-driven metal ion flux changes sign at the middle of the axis ($x = 7.5$ cm). However, the absolute value of the aluminum ion flux around the target is much larger than that near the wafer because of the higher density of Al^+ around the target which was shown in Figs. 4 and 6. The neutral fluxes are driven by the density gradient and temperature gradient (diffusion) when the convective flow is neglected; thus, it is easy to understand the aluminum atom flux behavior (Fig. 7b) by examining its density profile shown in Fig. 6b and the neutral temperature profile in Fig.

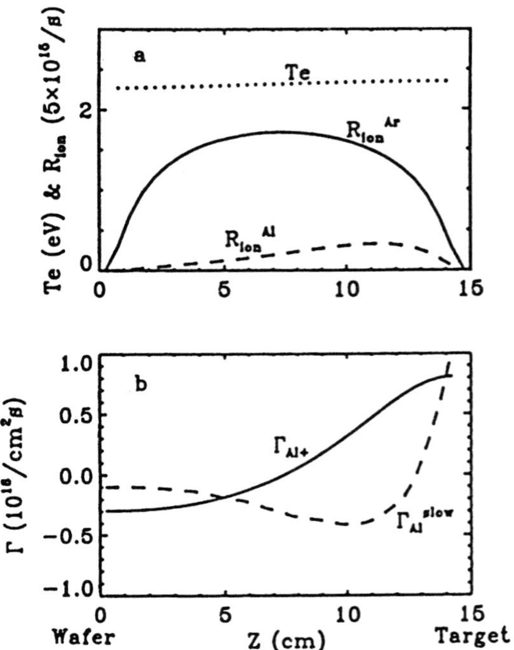

FIG. 7. 1-D profiles on chamber axis for (a) electron temperature T_e, ionization rate of argon and aluminum $R_{\text{ion}}^{\text{Ar}}$ and $R_{\text{ion}}^{\text{Al}}$; (b) fluxes of aluminum ions $\Gamma_{\text{Al+}}$ and thermalized aluminum atoms $\Gamma_{\text{Al}}^{\text{slow}}$.

8b. Because of the large gradient of aluminum atom density near the target, a large flux of neutral aluminum flows back to the target.

The background neutral argon atom density n_{Ar} and neutral temperature profile T_{neu} shown in Fig. 8 demonstrate an obvious neutral density decrease near the target which has been observed in sputtering experiments.[29,30,45] The model predicts that the neutral temperature can be as high as 800 K around the target by the intensive heating from fast sputtered atoms and plasma ions, and the temperature decreases to about 500 K near the wafer.

C. Comparison of Model Results to Experimental Measurements

Due to the relatively short time since the I-PVD system was developed, there are relatively few experimental measurements that can be used to compared with the simulations. In this section, we compare our simulation predictions to the available experimental data.[26,46–49]

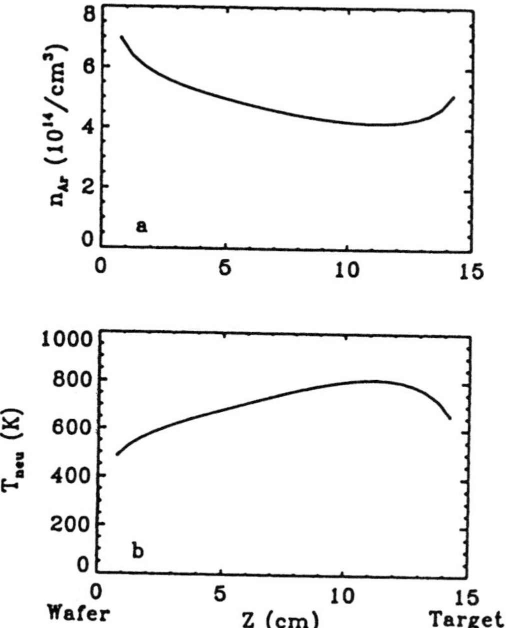

FIG. 8. 1-D profiles on chamber axis for (a) neutral argon atom density n_{Ar} and (b) neutral temperature.

1. Effects of RF Power

Figures 9 and 10 show the effects of the RF power on the plasma parameters and film deposition rate for three RF-inductive power levels: 500, 1000, and 1500 W. The neutral pressure is kept at 35 mTorr. Similar to pure argon inductively coupled discharges, Fig. 9a shows that the peak electron density increases linearly with the RF power applied. It is interesting that when the RF power increases, the volume-averaged electron density and argon ion density $\langle n_e \rangle$ and $\langle n_{Ar^+} \rangle$ increase linearly with power, whereas the metal ion density increases much more slowly (Fig. 9b). As a result, the ratio of $\langle n_{Al^+} \rangle / \langle n_{Ar^+} \rangle$ decreases as the RF power increases (Fig. 9c). These results are in qualitative agreement with the measurements of Nichols *et al.*[46] performed in an argon–copper I-PVD system. In their measurement, the electron density increases linearly with RF power; however, the metal ion density saturated at some RF power. Thus, the ratio of the metal ion density to the argon ion density decreases as power increases. The neutral temperature, both peak value and volume-averaged value shown in Fig. 9d, increase

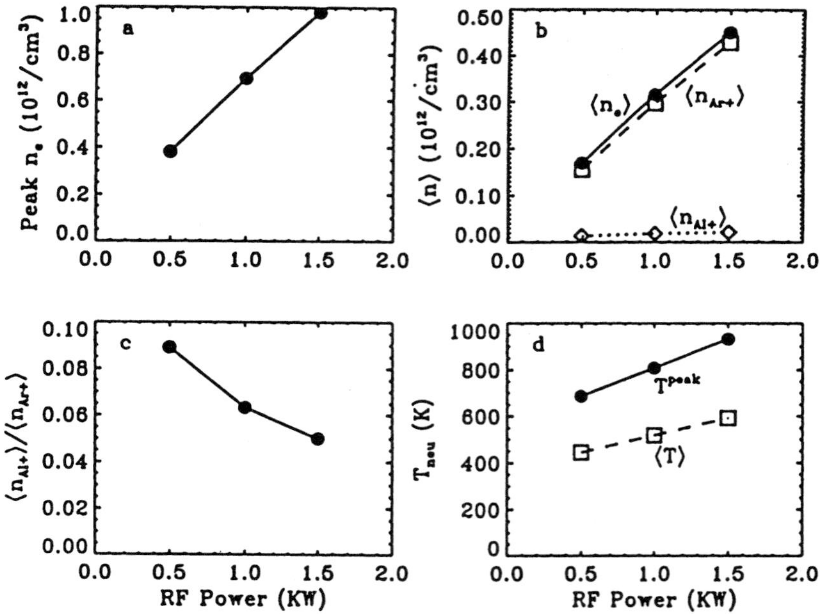

FIG. 9. Variations of (a) peak electron density n_e, (b) volume-averaged charged species densities $\langle n_e \rangle$, $\langle n_{Ar^+} \rangle$, and $\langle n_{Al^+} \rangle$, (c) ratio of volume-averaged metal ion density to argon ion $\langle n_{Al^+} \rangle / \langle n_{Ar^+} \rangle$, (d) peak and volume-averaged neutral temperature T^{peak} and $\langle T \rangle$ vs RF power deposited in the plasma.

linearly with RF power. The main reason is that the positive ion density, and therefore the ion current flowing to the target, increases with RF power. Thus, the total sputtered atom flux and the energy transfer rate to the background neutral atoms increase due to heating from the sputtered fast metal atoms. In addition, the heating from the fast ion species by charge exchange and collisions also increases with the ion density, which in turn increases linearly with RF power.

The average aluminum film deposition rate and the flux fraction of each deposition component [aluminum ions (Al^+) and thermalized and unthermalized aluminum atoms (Al^{slow} and Al^{fast}) are shown in Fig. 10. The total film deposition rate in Fig. 10a increases linearly with the applied RF power, which is consistent with the experimental measurements of Nichols *et al.*[46] One possible explanation is that the higher plasma density at higher RF power leads to a higher ionization degree of metal atoms, and then a larger metal ion flux (deposition rate) can be obtained on the substrate. However,

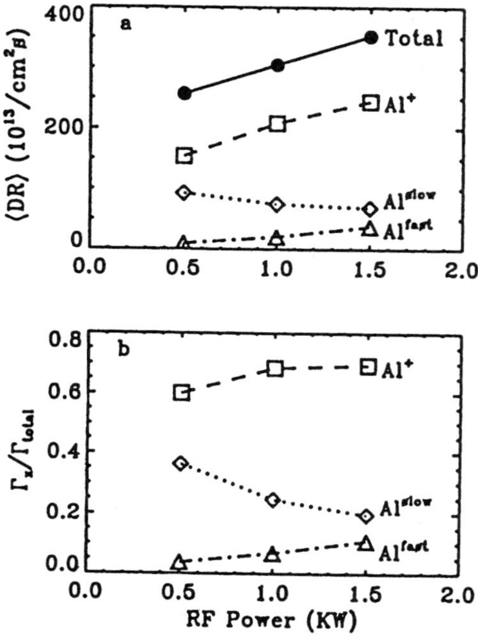

FIG. 10. Effects of RF power on (a) the average metal film deposition rate $\langle DR \rangle$ and (b) flux fraction of each film deposition component Al^+, Al^{slow}, and Al^{fast}.

the higher ionization degree of metal atoms at higher RF power tends to decrease the slow metal atom density near the substrate. Consequently, metal deposition from these species is reduced at higher RF power. Furthermore, the higher background neutral temperature and lower densities cause fewer collisions between the fast sputtered atoms and the background neutral atoms. Thus, the fast neutral thermalization process is less efficient than that for low RF power. As a result, the contribution of the unthermalized metal atoms on the film deposition is more important for high RF power cases. For example, the simulation shows that at 1500 W, the flux of unthermalized metal atoms can be as high as 10% of the total deposition flux. The ion flux fraction increases with RF power and saturates at about 70% when the power is higher than 1000 W for the geometry considered here. This result is also consistent with the experimental measurements[46–49] for copper film deposition. In these measurement, the metal–ion flux fraction saturated at 1000 W to a value of 85% for the pressure range of 30–36 mTorr.

2. *Electron Temperature Cooling Effect of Metal Species*

It has been suggested that a decrease in electron temperature and density occurs at high metal atom concentration in the plasma.[48] This is referred to as the "electron temperature cooling effect." However, recently the first direct measurement of the electron temperature cooling effect was done using a Langmuir probe in an aluminum I-PVD system (Fig. 11).[26] In the figure, the measured electron temperature is plotted as a function of magnetron current at two pressures: 10 and 30 mTorr. Note that Dickson *et al.*[26] show an equivalent Al atom density corresponding to each value of the magnetron current. Model predictions[50] of electron temperature were also plotted in the figure as a dashed line.

The comparison of the simulation predictions with the experimental measurements for electron temperature is given in Fig. 12a by neglecting the difference of chambers used for the simulation and experiments. The aluminum density shown in Fig. 11 is transformed as a fractional concentration $C_{Al}[C_{Al} = \langle n_{Al}\rangle/(\langle n_{Al}\rangle + \langle n_{Ar}\rangle)]$. Compared to the measurements,

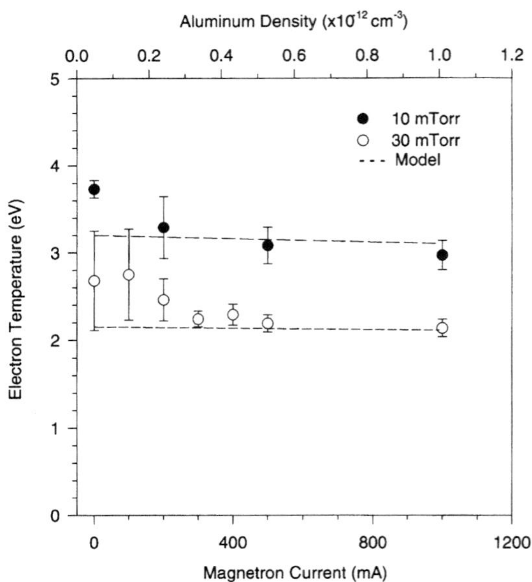

FIG. 11. The first measurement of electron temperature cooling effects of aluminum species in inductively coupled plasma. Global model results for electron temperature are also shown (dashed lines) (reprinted from Dickson *et al.*,[26] by permission of the American Vacuum Society).

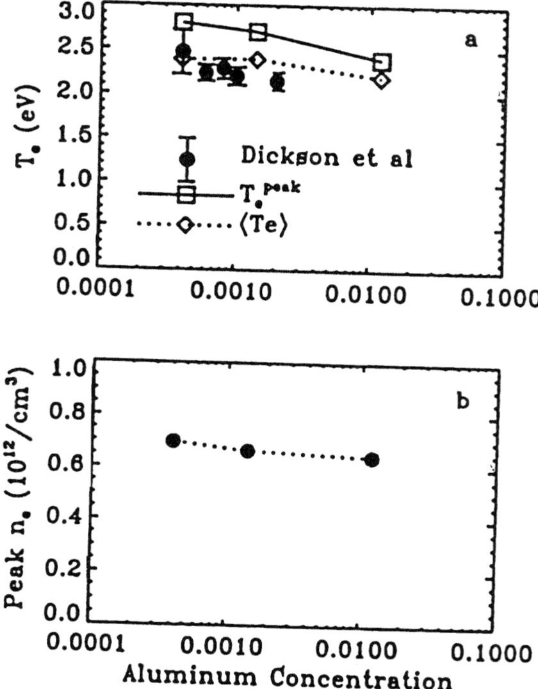

FIG. 12. Effects of metal species concentration on (a) electron temperature and (b) plasma density.

the simulation reproduces the trends of electron temperature variation when the concentration of aluminum C_{Al} increases. For the simulation results, within the regime of low metal concentration ($C_{\mathrm{Al}} < 0.2\%$), the electron temperature cooling effects are not dramatic. However, when C_{Al} increases to about 1%, the peak electron temperature decreases by 0.65 eV and the corresponding volume-averaged electron temperature decreases by 0.4 eV compared with the low metal concentration case ($C_{\mathrm{Al}} = 0.04\%$). However, this effect may not be observed in general because the increase of total sputtering flux tends to increase the background neutral temperature and this will decrease the neutral density. As a result, the electron temperature tends to increase. This electron temperature increase trend competes with the elebFctron temperature decrease trend due to the introduction of metal atoms. Therefore, the electron temperature could actually increase in some cases with an increase in metal concentration.

A slight decrease of the peak electron density is also predicted by the simulation when the metal species concentration increases, as shown in Fig. 12b. This may not always be true for other cases.

D. STUDIES OF FILM DEPOSITION RATE UNIFORMITY

Uniformity is one of the most important issues in thin film processing. The metal film deposition uniformity across the wafer in the I-PVD system is directly determined by the three components of metal species: metal ions, thermalized metal atoms, and unthermalized metal atoms. Uniformity is influenced by many possible factors, for example, the erosion patterns of target sputtering, the coil sputtering (determined by coil shape and location), the dimensions of the chamber, and the distance between target and the substrate. In this section, we have explored several effects on the film deposition uniformity. Unfortunately, no experimental measurements are available to compare to the predictions.

1. Effects of Target Erosion Patterns

The target is of course eroded by the impact of high-energy ions, and the target erosion pattern is the result of the net sputtering flux distribution of metal atoms along the target. Generally, the target is the main source of fast metal atoms.

From Fig. 5, it is obvious that the radial nonuniformity of the film deposition rate is mainly caused by the radially nonuniform distribution of metal ion flux. The on-axis peak of metal ion density leads to a higher film deposition rate at the center of substrate. Here, we try to obtain a more uniform film deposition by changing the transport of neutral metal in the chamber in order to move the peak of the metal ion density off the chamber axis. Thus, a kind of target erosion pattern called the "ring target" is chosen. This means that the target is sputtered only in an annular ring defined by $5 < r < 17.5$ cm. By comparison, the target erosion pattern used in the previous simulations is referred to as "disk target" because the metal atoms are sputtered through the whole area of the target from $r = 0$ to $r = 20$ cm. The operating conditions used here are the same as those used in Fig. 4.

The neutral and charged species density and temperature profiles, the plasma potential, plasma power deposition, and the heat exchange between the fast atoms and the background neutrals are shown in Fig. 13 for the ring target case. Compared to Fig. 4, it is obvious that for such a low metal species concentration ($<0.05\%$), the effects of the metal sputtering profile on the plasma potential, power deposition, electron density, and electron temperature are negligible. However, the neutral density and temperature

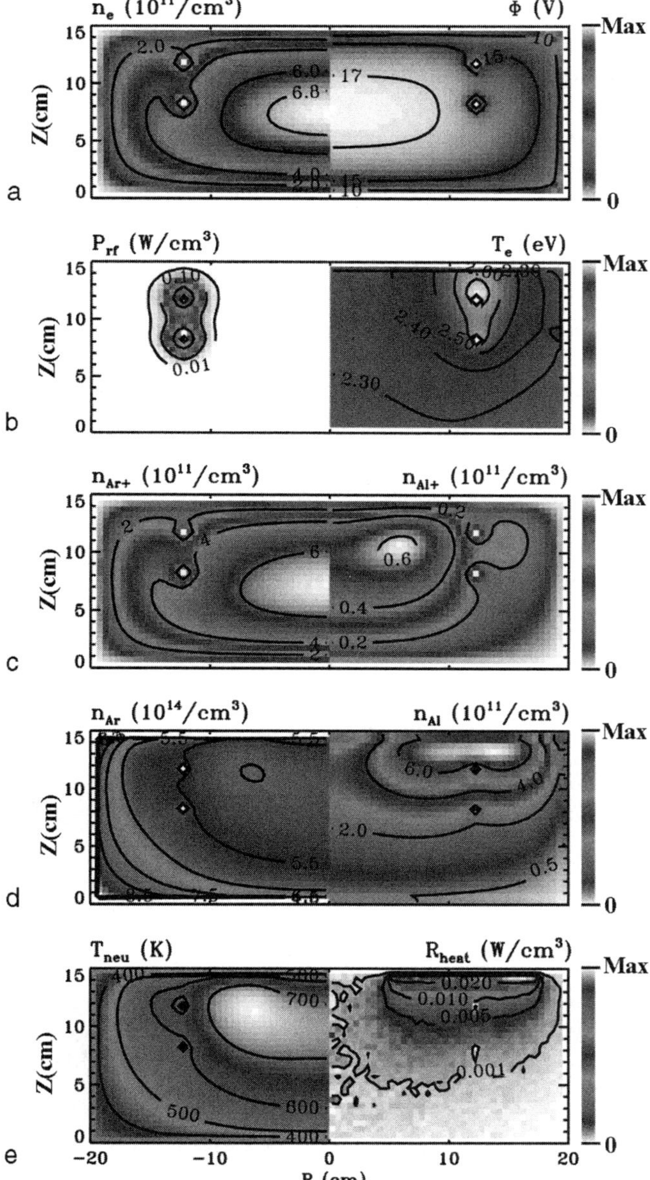

FIG. 13. Profiles of (a) electron density n_e and plasma potential Φ, (b) RF power deposition P_{rf} and electron temperature T_e, (c) ion densities n_{Ar^+} and n_{Al^+}, (d) neutral densities Ar and Al, and (e) neutral temperature T_{neu} and the energy transfer rate R_{heat} from fast atoms to slow background gases with ring target erosion pattern.

profiles are affected. The fast metal atom thermalization rate around the target for the ring target is compared with the profile for the disk target sputtering in Fig. 14. The thermalized aluminum atom peaks at approximately $r = 10$ cm and a few centimeters below the target. In addition, the peak of the aluminum ion density moves out radially to about 6 cm from the chamber axis. Similarly, because of the high energy transfer rate from fast atoms to slow atoms under the ring target shown in Figs. 13 and 14, the peak of neutral temperature also moves off axis.

The total aluminum film deposition rate and the fraction of aluminum ions in the film deposition for disk target and ring target are compared in Fig. 15. Due to the off-axis metal ion and atom density profiles, the film deposition is considerably more uniform with the ring target. In addition, due to the similar volume-averaged neutral temperature and background neutral density, the density of the thermalized metal near the substrate is similar for both sputtering profiles. Thus, the average film deposition rate across the substrate is similar in the two cases. Furthermore, because the different sputtering flux patterns on the target mainly influence the profiles of metal species in the radial direction and the uniformity of film deposition, the fraction of the ion flux does not change much for the disk target and ring target erosion patterns.

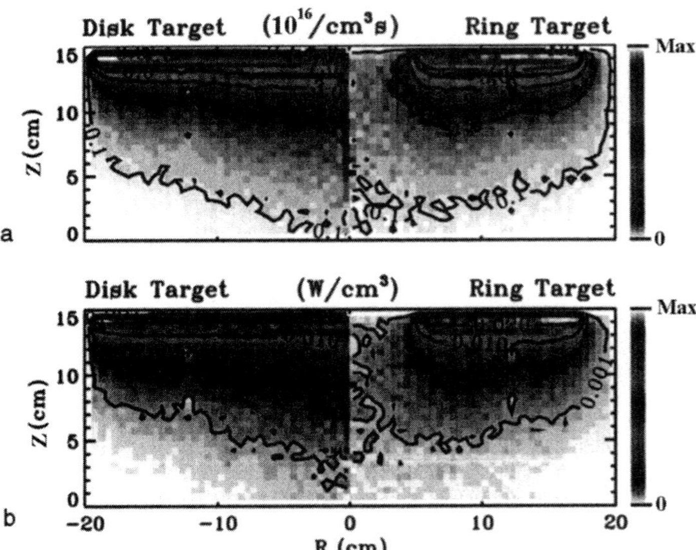

FIG. 14. Comparisons of fast aluminum atom (a) thermalization profiles and (b) background neutral atom heating with disk target and ring target erosion patterns.

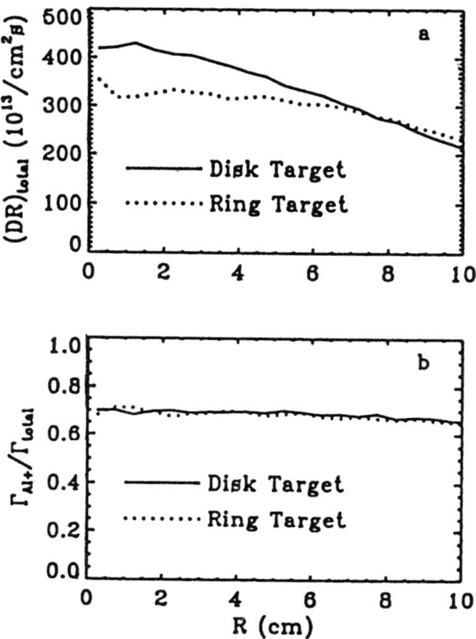

FIG. 15. (a) Total film deposition rate and (b) metal ion flux fraction for disk target and ring target erosion patterns.

2. *Effects of Coil Sputtering*

Due to the high voltage produced on the RF coil, the coil can be sputtered by the impact of energetic ions. The film deposition rate uniformity on the substrate can be affected by the coils. For example, the voltage on the coil, the relative location of the coil with respect to the substrate, the coil shape, and the coil surface area could influence deposition rate uniformity. We have examined only a few of the possible effects in the results we report here.

a. General Effects of Coil Sputtering Generally, when the RF coil is immersed in the plasma, sputtering from the coil causes an increase in the total sputtered flux of metal species entering the chamber. As a result, more energy is transfered to the slow background gas atoms and the neutral temperature tends to increase. Simultaneously, since the RF coil is generally located off the chamber axis, the related radial profiles tend to be changed, particularly for metal species. However, the magnitude of these influences will depend on the specific operating conditions.

The 2-D plasma profiles and aluminum film deposition characteristics with coil sputtering are shown in Figs. 16 and 17, respectively. The operating conditions are the same as those used for Figs. 4 and 5 except for the additional coil sputtering. The coil sputtering fluxes are assumed to be the total ion flux arriving at the coil surfaces multiplied by the coil sputtering

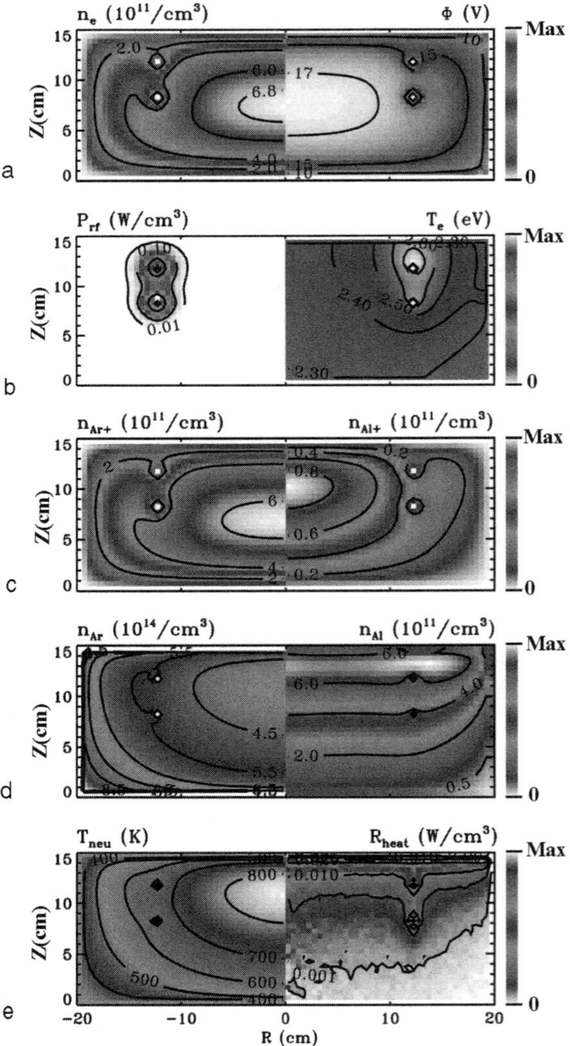

FIG. 16. 2-D plasma profiles with coil sputtering. The operating conditions are the same as those in Fig. 4 except for the addition of coil sputtering.

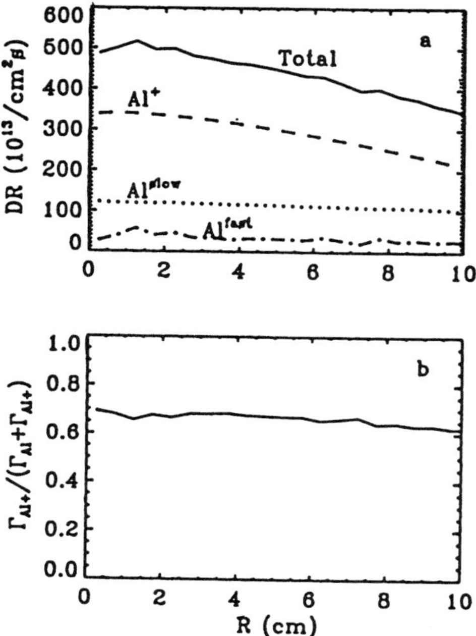

FIG. 17. (a) Total aluminum film deposition rate and contributions from each component of aluminum ions (Al^+), thermalized aluminum atoms (Al^{slow}), and unthermalized aluminum atoms (Al^{fast}). (b) Flux ratio of aluminum ions (Al^+) to the total deposition flux across the substrate. The operating conditions are same as those in Fig. 4, except for the addition of coil sputtering.

yield (Y_{coil}, assumed to be 1). This implies that the average energy of ions on the coil during the RF period is about 500 eV. It is apparent that there are no large difference for the plasma profiles with and without coil sputtering, except for the metal species. Because the coil surface area is smaller than that of the target, the concentration of the metal species in the plasma does not increase significantly when sputtering from coils is added. However, due to the higher energy transfer rate from the unthermalized metal atoms around the RF coil, the background neutral temperature increases locally and the argon atom density decreases slightly. At the same time, because of the slight increase of the total amount of sputtered aluminum in the chamber, the thermalized aluminum atom density and aluminum ion density also increase slightly. As mentioned previously, sputtering from the coil represents an off-axis source of aluminum. At the substrate, this results in more uniform Al atom and ion density profiles compared to the case without coil sputtering (Fig. 4).

Compared with Fig. 5, the total film deposition rate increases a few percent due to the addition of coil sputtering. All the metal species flux profiles across the substrate, both ions and neutrals, are more uniform when the coil sputtering is included. The ratio of film deposition rate at the edge of substrate to that at the substrate center increases to 70% from 50% with coil sputtering. Similar to the case without coil sputtering, the nonuniformity of the film deposition is still mainly caused by the radially nonuniform ion distribution. In addition, the metal ion fraction of the total metal flux has decreased with coil sputtering because the neutral metal flux has increased more than the metal ion flux.

b. Effects of Coil Location The distance from the metal sputtering source to the substrate is an important parameter affecting the thermalization process of the fast metal component. When the coil sputtering is significant, the location of coils with respect to the substrate could be important for film deposition on the substrate. In order to see clearly the effects of coil sputtering, the metal sputtering from the top target is turned off, and only the coil sputtering is considered in the following results.

The results for two cases with different coil locations are demonstrated here. One has the same coil configuration as in Fig. 3. The minimum distance from the lower turn of coil to the chamber bottom where the substrate is located is 8 cm. We refer to this as the "upper coils" configuration. The second coil configuration is referred to as "lower coils" configuration. In this configuration, the minimum distance from the lower turn of coil to the chamber bottom is decreased to 4 cm. In both cases, the RF power and neutral pressure are the same as those in the case shown in Fig. 4: 1000 W and 35 mTorr, respectively. The coil sputtering yield is again assumed to be 1.

As expected, the effects of the coil sputtering are mainly on the metal species, with no noticeable effect on the other plasma profiles. As shown in Fig. 18, the 2-D profiles are compared for upper coils (left) and lower coils (right) cases. The electron density profiles for both cases are very similar. However, since the coil sputtering is the only source of fast metal atoms in the current simulation, the fast metal species thermalization occurs mostly around the coil. The densities of aluminum ions and thermalized aluminum atoms are peaked correspondingly near the coils.

The aluminum film deposition rate and aluminum ion flux fraction for these two cases are shown in Fig. 19. The off-axis sputtering profile of fast metal atoms results in a deposition rate profile that peaks at the substrate edge. Especially for the lower coils case, this trend is stronger due to the large contributions from the thermalized and unthermalized metal atoms. Simultaneously, because of the short distance from the coils to the outer edge of the substrate, more neutral metal arrives there before being ionized

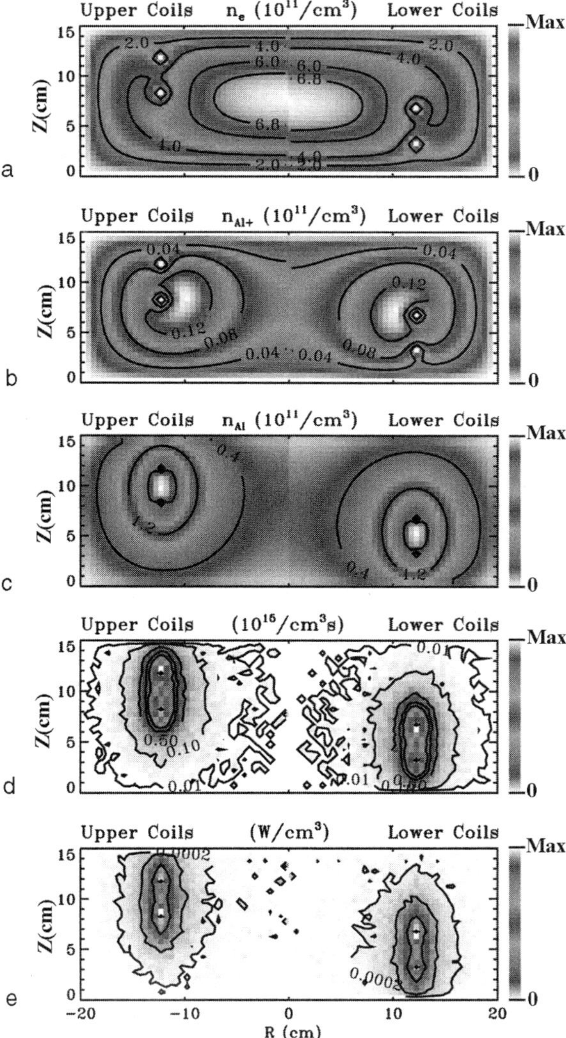

FIG. 18. 2-D plasma profiles for pure coil sputtering with different coil locations. The same RF power (1000 W), pressure (35 mTorr), and coil sputtering yield ($Y_{\text{coil}} = 1$) as those used in Fig. 4 are used here.

in the plasma in the lower coils case. Thus, a lower metal ion flux fraction is obtained in this region.

Of course, in an actual I-PVD tool, the coil sputtering and target sputtering exist simultaneously. Recalling the previous results, the thermalization process of fast species is greatly affected by the background neutral

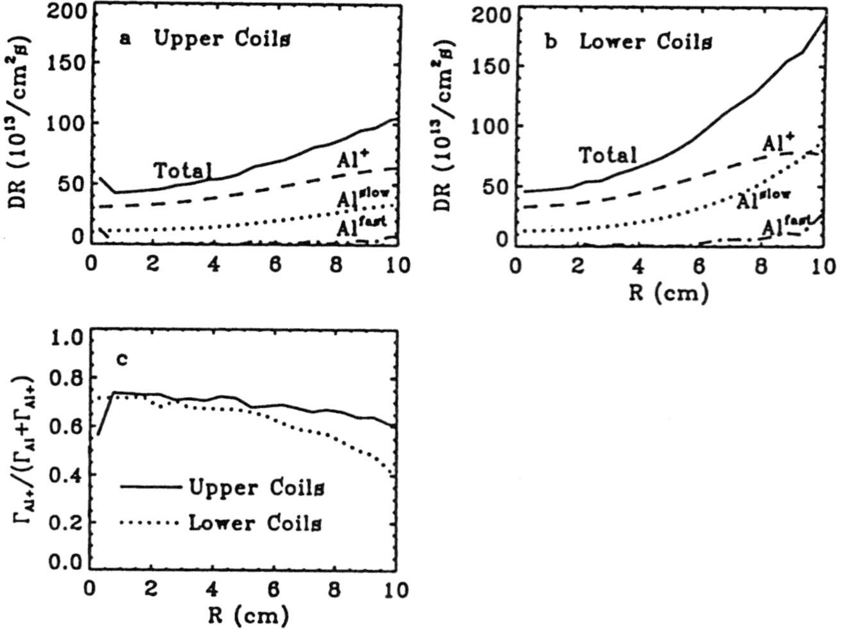

FIG. 19. Aluminum film deposition rate for pure coil sputtering with different coil locations of (a) upper coils and (b) lower coils; (c) aluminum ion flux fraction.

density, and the neutral density is determined by the pressure and neutral temperature. However, the neutral temperature is influenced most by the heat transfer rate from fast neutrals during the thermalization process. This coupling means that the effects of coil sputtering in real processing cannot truly be investigated separately from the target sputtering. The importance of coil sputtering will be related to the target sputtering characteristics, including the total sputtering flux from the target, the location of the target, and the target erosion pattern.

3. *Effects of Other Factors*

There are many other factors that could influence the film deposition and species transport for I-PVD systems (e.g., chamber geometry and the background gas pressure). These effects can be simulated by the model described in this chapter, and these issues are left for future studies.

E. FEATURE PROFILE SIMULATION OF METAL FILM DEPOSITION

In this section, we present an example of a profile evolution for a 0.5-μm trench filling by aluminum. This simulation is based on the profile simulation model given in Section II,E. We assume no bias is applied on the substrate, so the impacting energy of ions is relatively low. This suggests that the film resputtering or film etching is negligible. Furthermore, a unity sticking coefficient is assumed for all metal components (ions and thermalized and unthermalized atoms), and no surface reflection or surface diffusion are considered for the arriving metal species.

The angular distribution functions (ADFs) for the different components of the metal species are shown in Fig. 20. The incident flux for each metal component and the angular distributions for unthermalized aluminum atoms and aluminum ions are obtained from the plasma simulations corresponding to the conditions shown in Fig. 4. The thermalized aluminum atom flux is assumed to be completely randomized and its ADF is therefore a cosine distribution, as shown in Fig. 20a by the dotted line. The ADF for

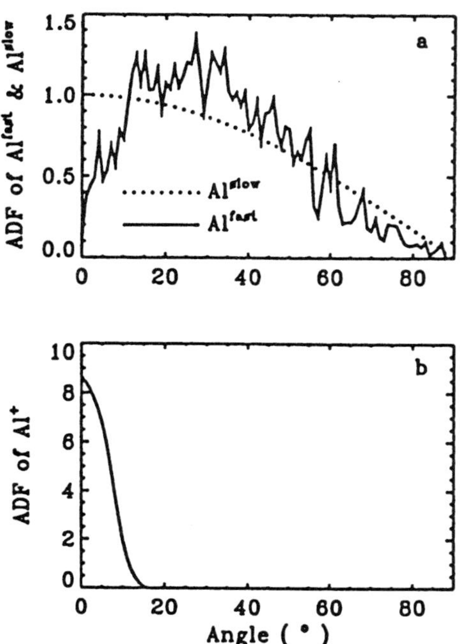

FIG. 20. Angular distributions on the substrate surface for (a) thermalized and unthermalized aluminum atoms and (b) aluminum ions without bias applied.

the unthermalized aluminum is obtained from the neutral fast-particle Monte Carlo simulation described in Section II,B. It is interesting that the ADF of unthermalized atoms, initiated from an isotropic distribution on the target, tends to peak at around 20° from the substrate normal after experiencing a series of collisions with the slow background neutral atoms. The ADF for fast Al atoms was obtained by averaging results at the center of the substrate with an effective diameter of 2 cm. For filling high aspect ratio features, this kind of angular distribution tends to contribute more to the overhang structure around the top corner of the trenches compared to the isotropic distribution because of the relatively small flux in the normal direction. The ADF for aluminum ions is calculated from the ion Monte Carlo simulation for the bulk plasma in combination with an analytical sheath model as described in Section II,D. As shown in Fig. 20b, a typically narrow angular distribution for ions is obtained, and most of the ions impact the substrate with a small angle from the normal direction ($\leqslant 8 \sim 10^\circ$).

The profile evolution of aluminum film deposition for a 0.5-μm trench with aspect ratio 2:1 is shown in Fig. 21. Figure 22 is a SEM of Al trench

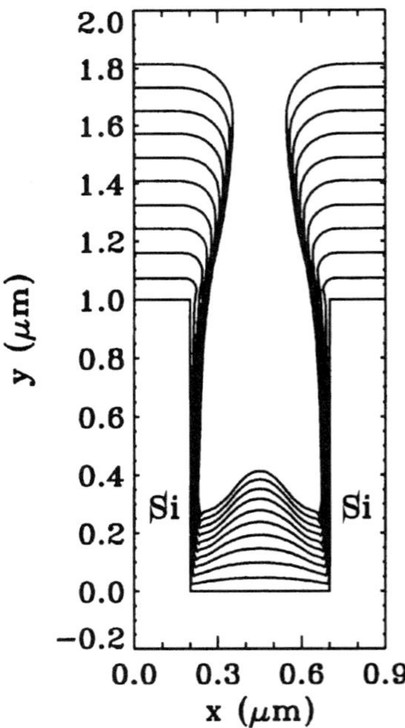

FIG. 21. Profile evolution of aluminum film deposition for a 0.5-μm trench with aspect ratio 2:1.

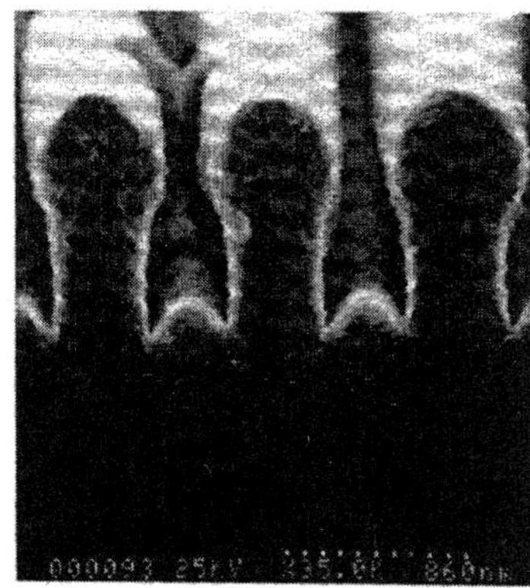

FIG. 22. SEM pictures of Al deposition into trenches with aspect ratio 2.0. The ion to neutral flux ratio is approximately 1:1 (reprinted from Hamaguchi and Rossnagel,[51] by permission of the American Vacuum Society).

filling under conditions similar to those used in the simulation results shown in Fig. 21. Topographically, the modeling results appear to be in fairly good agreement with the experiments. However, we do not claim that the apparently good agreement between the simulation prediction and the SEM represents a serious test of the model. The agreement suggests that the basic simulation methodology is reasonable. Much more thorough comparisons between the simulation predictions and measured profiles are needed to refine and extend the model.

Acknowledgments

The authors are grateful for partial support from Applied Material Lam Research, and the California State MICRO program.

References

1. J. D. Bukowski, D. B. Graves, and P. Vitello. *J. Appl. Phys.*, **80**, 2614 (1996).
2. D. B. Graves and K. F. Jensen, *IEEE Trans. Plasma Sci.* **PS-14**, 78 (1986).
3. D. B. Graves, *J. Appl. Phys.* **62**, 88 (1987).
4. J.-P. Boeuf, *Phys. Rev. A* **36**, 2782 (1987).
5. E. Gogolides, J.-P. Nicolai, and H. H. Sawin, *J. Vac. Sci. Technol. A* **7**, 1001 (1989).
6. G. R. Misium, A. J. Lichtenberg, and M. A. Lieberman, *J. Vac. Sci. Technol. A* **7**, 1007 (1989).
7. M. Meyyappan and J. P. Kreskovsky, *J. Appl. Phys.* **68**, 1506 (1990).
8. S.-K. Park and D. J. Economou, *J. Appl. Phys.* **68**, 3904 (1990).

9. M. Meyyappan and T. R. Govindan, *J. Vac. Sci. Technol. A* **10**, 1344 (1992).
10. Y. Yasaka, A. Fukuyama, A. Hatta, and R. Itatani, *J. Appl. Phys.* **72**, 2652 (1992).
11. T. J. Sommerer and M. J. Kushner, *J. Appl. Phys.* **71**, 1654 (1992).
12. M. J. Kushner, *J. Appl. Phys.* **71**, 4173 (1992).
13. R. K. Porteous and D. B. Graves, *IEEE Trans. Plasma Sci.* **19**, 204 (1991).
14. P. L. G. Ventzek, T. J. Sommerer, R. J. Hoekstra, and M. J. Kushner, *Appl. Phys. Lett.* **63**, 605 (1993).
15. J. D. Bukowski, R. A. Stewart, D. B. Graves, and P. Vitello, in *Proceedings of the 10th Symposium on Plasma Processing* (G. S. Mathad and D. W. Hess, Eds.), Vol. 94-20, pp. 87–96, Electrochemical Society, Pennington, NJ, 1994.
16. R. S. Wise, D. P. Lymberopoulos, and D. J. Economou, *Plasma Sources Sci. Technol.* **4**, 317 (1995).
17. M. Meyyappan and T. R. Govindan, *J. Appl. Phys.* **78**, 6432 (1995).
18. C. Lee, D. B. Graves, and M. A. Lieberman, *Plasma Chem. Plasma Process.* **16**, 99 (1996).
19. M. Li, H. Date, and D. B. Graves, in *Electron Kinetics and Applications of Glow Discharges* (U. Kortshagen and T. D. Tsendin, Eds.), Plenum, New York, 1998.
20. M. J. Grapperhaus, Z. Krivokapic, and M. J. Kushner, *J. Appl. Phys.* **83**, 35 (1998).
21. R. A. Stewart, P. Vitello, D. B. Graves, E. F. Jaeger, and L. A. Berry, *Plasma Sources Sci. Technol.* **4**, 36 (1995).
22. M. M. Turner, *Phys. Rev. Lett.* **71**, 1844 (1993).
23. E. F. Jaeger, L. A. Berry, J. S. Tolliver, and D. B. Batchelor, *Phys. Plasmas* **2**, 2597 (1995).
24. L. R. Peterson and J. E. Allen, Jr., *J. Chem. Phys.* **56**, 6068 (1972).
25. L. L. Shimon, E. I. Nepiipov, and I. P. Zapesochnyi, *Sov. Phys. Tech. Phys.* **20**, 434 (1975).
26. M. Dickson, F. Qian, and J. Hopwood, *J. Vac. Sci. Technol. A* **15**, 340 (1997).
27. J. A. Valles-Abarca and A. Gras-Marti, *J. Appl. Phys.* **55**, 1370 (1984).
28. V. V. Serikov and K. Nanbu, *J. Vac. Sci. Technol. A* **14**, 3108 (1996).
29. D. W. Hoffman, *J. Vac. Sci. Technol. A* **3**, 561 (1985).
30. S. M. Rossnagel, *J. Vac. Sci. Technol. A* **6**, 19 (1988).
31. A. Bogaerts, M. V. Straaten, and R. Gijbels, *J. Appl. Phys.* **77**, 1868 (1995).
32. G. M. Turner, *J. Vac. Sci. Technol. A* **13**, 2161 (1995).
33. Y. Yamamura and M. Ishida, *J. Vac. Sci. Technol. A* **13**, 101 (1995).
34. H. M. Urbassek and D. Sibold, *J. Vac. Sci. Technol. A* **11**, 676 (1993).
35. A. Kersch, W. Morkoff, and C. Werner, *J. Appl. Phys.* **75**, 2278 (1994).
36. F. L. Tabares and D. Tafalla, *J. Vac. Sci. Technol. A* **14**, 3087 (1996).
37. J.-P. Boeuf and E. Marode, *J. Phys. D Appl. Phys.* **15**, 2169 (1982).
38. M. Surendra, D. B. Graves, and G. M. Jellum, *Phys. Rev. A* **41**, 1112 (1990).
39. E. W. McDaniel, in *Atomic Collisions: Electrons and Photon Projectiles*, Wiley, New York, 1989.
40. J. B. Hasted, in *Physics of Atomic Collisions*, Butterworth, London, 1964.
41. V. A. Godyak and N. Sternberg, *Phys. Rev. A* **42**, 2299 (1990).
42. M. Dalvie, R. T. Farouki, and S. Hamaguchi, *IEEE Trans. Electron Devices* **39**, 1090 (1992).
43. M. Tuda, K. Nishikawa, and K. Ono, *J. Appl. Phys.* **81**, 960 (1997).
44. J. Ignacio, F. Ulacia, and J. P. McVittie, *J. Appl. Phys.* **65**, 1484 (1989).
45. S. M. Rossnagel, submitted for publication.
46. C. A. Nichols, S. M. Rossnagel, and S. Hamaguchi, *J. Vac. Sci. Technol. B* **14**, 3270 (1996).
47. S. M. Rossnagel and J. Hopwood, *Appl. Phys. Lett.* **63** 3285 (1993).
48. S. M. Rossnagel and J. Hopwood, *J. Vac. Sci. Technol. B* **12**, 449 (1994).
49. P. F. Cheng, S. M. Rossnagel, and D. N. Ruzic, *J. Vac. Sci. Technol. B* **13**, 203 (1995).
50. J. Hopwood and F. Qian, *J. Appl. Phys.* **78**, 758 (1995).
51. S. Hamaguchi and S. M. Rossnagel, *J. Vac. Sci. Technol. B* **13**, 183 (1995).

Subject Index

Recent Volumes In This Series

Maurice H. Francombe and John L. Vossen, *Physics of Thin Films*, Volume 16, 1992.

Maurice H. Francombe and John L. Vossen, *Physics of Thin Films*, Volume 17, 1993.

Maurice H. Francombe and John L. Vossen, *Physics of Thin Films, Advances in Research and Development, Plasma Sources for Thin Film Deposition and Etching*, Volume 18, 1994.

K. Vedam (guest editor), *Physics of Thin Films, Advances in Research and Development, Optical Characterization of Real Surfaces and Films*, Volume 19, 1994.

Abraham Ulman, *Thin Films, Organic Thin Films and Surfaces: Directions for the Nineties*, Volume 20, 1995.

Maurice H. Francombe and John L. Vossen, *Homojunction and Quantum-Well Infrared Detectors*, Volume 21, 1995.

Stephen Rossnagel and Abraham Ulman, *Modeling of Film Deposition for Microelectronic Applications*, Volume 22, 1996.

Maurice H. Francombe and John L. Vossen, *Advances in Research and Development*, Volume 23, 1998.

Abraham Ulman, *Self-Assembled Monolayers of Thiols*, Volume 24, 1998.

Subject and Author Cumulative Index, Volumes 1–24, 1998.

Ronald A. Powell and Stephen Rossnagel, *PVD for Microelectronics: Sputter Deposition Applied to Semiconductor Manufacturing*, Volume 26, 1998.

ISBN 0-12-533027-8